BIOCHEMISTRY RESEARCH TRENDS

BIOCHEMISTRY AND MOLECULAR BIOLOGY COMING OF AGE

BIOCHEMISTRY RESEARCH TRENDS

Additional books in this series can be found on Nova's website under the Series tab.

Additional e-books in this series can be found on Nova's website under the e-book tab.

BIOCHEMISTRY RESEARCH TRENDS

BIOCHEMISTRY AND MOLECULAR BIOLOGY COMING OF AGE

MARY MIU YEE WAYE
EDITOR

nova publishers
New York

For permission to use material from this book please contact us:
Telephone 631-231-7269; Fax 631-231-8175
Web Site: http://www.novapublishers.com

Library of Congress Cataloging-in-Publication Data

ISBN: 978-1-63321-815-4

Library of Congress Control Number: 2014948048

Published by Nova Science Publishers, Inc. † New York

CONTENTS

EDITORIAL

BIOCHEMISTRY AND MOLECULAR BIOLOGY IN HONG KONG: PAST AND FUTURE

*T. R. C. Boyde**
Emeritus Professor of Biochemistry,
University of Hong Kong

When Frederick Gowland Hopkins [1] attended the City of London School, seventy years before me, the word biochemistry did not exist in English. Even when he was invited to Cambridge by Michael Foster [2] it was to teach "the chemical side of physiology" though within a few years Biochemistry there had broken away from its parent, leading the way for many departments like it because chemically-minded people wished to shed their physiological strait-jacket. Cambridge took the plunge in 1914, the University of Hong Kong (HKU) [3] not until 1965 and Professor Walshe arrived the following year. At Chinese University of Hong Kong (CUHK) [4] the alternate route was followed. Since no medical school yet existed [5] Biochemistry began in science faculty, and by 1973 the department under Professor Ma Lin [6] was far better developed than HKU. They had all the things we lacked, though there was no clear route to research funding for CUHK people or anyone else in Hong Kong. There was no Croucher Foundation [7], no research grant committees (such as GRF funds from the University Grants Council or UGC-8, ITF funds from the Innovation and Technology Commission or ITC- 9, or RFCID/HHRSF funds from the Food and Health Bureau-10), nothing. It all seems like a century ago.

Ted Walshe's years at HKU were frustrating in many ways. He started a program in science faculty, generating resentment which still endures among those who felt that they were losing students rather than gaining a partner; and was once heard to say, "If they want a — mess they can have a — mess", where 'they' means the Faculty of Medicine: so bad treatment by Faculties was reassuringly even-handed. To be fair, times were still tough in Hong Kong; no-one wanted to invest in basic science and it showed. On my first visit to the HKU biochemistry department, in 1972 after Ted's early death, the words that sprang to mind were, "marble halls": there was no-one at the benches and nothing going on. One room

* E-mail: tom@boyde.com.

designated as the "Research Laboratory" was the demonstrators' domain; there was a cold room and an ultracentrifuge but scarcely any other functional modern equipment, the best spectrophotometer having been damaged some time earlier by an acid spill and never repaired; no bench for me to use until a new one was built. Typical of those bad old days, a second ultracentrifuge lay hidden elsewhere in the Li Shu Fan building [11]; bricked up when the lecturer who owned it left the place and rediscovered twenty years later. (We did put some of the rotors into use but the drive was obsolete and beyond repair.)

That's the raw truth about Biochemistry in Hong Kong, just over 30 years ago. It's hard to imagine now with all the sophisticated new buildings and advanced research in progress everywhere, but then it was a desperate matter of survival to raise the profile of the science, stimulate interest generally in doing research or paying for it to be done, and provide for cooperation and communication between the perfectly sound biochemists who were already in post. That is why the old Hong Kong Biochemical Association (now renamed as the Hong Kong Society of Biochemistry and Molecular Biology -12) came into existence, the final spark coming from Takashi Murachi [13] of Kyoto during a flying visit in 1977, "Tom, maybe you're right not to push, but there's nothing wrong with giving the flywheel a good spin."

So we did. The Association held its first conference, entitled "Science for a Better Life", in December 1978 (Figure 1). It was essentially a propaganda exercise directed at the general public. 200 people attended lectures in the old Chemistry Theatre of HKU and even today the papers look incredibly up-to-date and relevant. We had Lester Packer talking about bio-mass as a fuel source (carbon footprint before it became popular), Brian Morton on conservation of the sea shore, and so on. Winding up, I called attention to the need for new approaches to research funding in Hong Kong,

"To achieve this .... requires a reserve fund, independent of government or of the individual researcher himself; in other words a Research Foundation. No such body exists in Hong Kong today, in one of the richest countries of the world." (Contributions of the Hong Kong Biochemical Association, vol 1, no.1, page 77).

It took ten more years to get there and by roundabout ways. The Shirley Boyde Trust [14] made its first research grants in 1989, quickly followed by the Croucher Foundation, but still barely more than 20 years ago.

Those stories abut how biochemistry came to Hong Kong are all very well but maybe the future is more interesting, and to talk about that depends on understanding what biochemistry and molecular biology actually are. The latter phrase is used by many people to indicate that what they do is exciting biology and not dull old chemistry; again it seems to them an escape from a strait-jacket; biochemists were the little grey men obstructing progress. But the very words 'molecular biology' had been used for twenty years before the discoveries of Watson and Crick [15], to mean biochemistry and nothing else. It does not belittle the great contributions of those who wish to call themselves molecular biologists if we gently point out that means to sequence polypeptides and nucleic acids were both invented by a Cambridge biochemist of the old school, and similarly with most of the other fundamental advances.

Peter Campbell was a good friend of ours and he thought that biochemistry as an independent subject was no longer necessary, its job was done. This is to understand biochemistry as a servant of biology, which many biologists would like it to be, and as a finite body of knowledge which perhaps says more about Campbell as a research worker than anything else. Ted Walshe had a clearer idea of the subject; his concern was with 'the chemical phenomenon that we call Life'. My only quibble there is that he has *defined* life as a chemical phenomenon, an axiom rather than the thing required to be demonstrated. We might find it by experiment to be something else!

One thing that causes a lot of trouble is to equate biochemistry with biological chemistry, i.e. the study of biological compounds as objects of chemical interest. 'Organic chemistry' originally meant the chemistry of living organisms so it sounds like the same thing as biochemistry, but the phrase was then extended to all compounds of carbon. Today, there is hardly any resemblance between the work of an organic chemist and a biochemist, and when a biochemist does some synthetic work of his own he is acting as an organic chemist, a displaced person.

Suddenly everything changes when it comes to structural and physical studies of molecules having biological interest; so a better definition of biochemistry might be 'the application of chemical disciplines to the study of living organisms'. We should be thinking chemically but in a biological environment. Included must be any kind of chemistry and perhaps physical aspects are more important than the others.

Is that just playing with words? We who are true biochemists don't think so and we may wish to exclude those who apply chemical methods but not chemical thinking to their research. To determine the nucleic acid sequence of an organism is, nowadays, either biology or chemistry, not biochemistry. Our interest begins when we wish to find out how a segment of nucleic acid is prepared for action in gene expression; why one species differs from another; why so much of human DNA is junk and does nothing (– a likely story, by the bye); or the cause of cancer.

It seems to me that biochemistry/molecular biology will continue to exist as an independent discipline for as long as science itself and be defined by two things; the nature of the subject matter and the basis of the approach to studying it; not chemical methods only, nor only applied objectives, but Life as a proper subject and chemical thinking as allowing an approach peculiarly its own. And Hong Kong may have a very important part to play in that: we oldsters certainly struggled to make it possible and hope that the effort was not in vain.

REFERENCES

[1] "Sir Frederick Hopkins - Biography". Nobelprize.org. 24 Aug 2010. Available from: http://nobelprize.org/nobel_prizes/medicine/laureates/1929/hopkins-bio.html

[2] "Michael Foster- Biography". Availabe from: http://vlp.mpiwg-berlin.mpg.de/references?id=per75

[3] The University of Hong Kong. Available from: http://www.hku.hk/

[4] The Chinese University of Hong Kong. Available from: http://www.cuhk.edu.hk/english/index.html

[5] The principal clinical teaching facility of the Medical Faculty of the Chinese University of Hong Kong, did not open until 1984. Available from: http://www.med.cuhk.edu.hk/v7/p5_15.htm

[6] "Professor Ma Lin- Biography" Press release of Shaw College, The Chinese University of Hong Kong. Available from: http://www.cuhk.edu.hk/ipro/pressrelease/ma_lin-e.htm

[7] The Croucher Foundation. Available from: http://www.croucher.org.hk/

[8] The GRF funds from University Grants Council (UGC) of Hong Kong. Available from: http://www.ugc.edu.hk/eng/rgc/grf/grf.htm

[9] Innovation and Technology Commission. Available from: http://www.itc.gov.hk/en/funding/itf.htm

[10] RFCID from the Food and Health Bureau. Available from: http://www.fhb.gov.hk/grants/English/funds/funds_rfcid/funds_rfcid_abt/funds_rfcid_abt.html and HHSRF from the Food and Health Bureau. Available from: http://www.fhb.gov.hk/grants/english/funds/funds_hhsrf/funds_hhsrf_abt/funds_hhsrf_abt.html

[11] "Li Shu Fan - Biography". Available from: http://en.wikipedia.org/wiki/Li_Shu_Fan

[12] The Hong Kong Society of Biochemistry and Molecular Biology. Available from: http://www.bch.cuhk.edu.hk/hksbmb/

[13] "Takashi Murachi-Obituary". Available from: http://informahealthcare.com/doi/pdf/10.3109/14756369109030391

[14] The Shirley Boyde Trust Lectures. Available from: http://www.hku.hk/biochem/news/lectures.html and Blog in Chinese Available from: http://blog.xuite.net/a220501/blog/18989807

[15] 'Secret of Life as it turns 50' by Ivan Noble. Updated 27 February 2003. Available from: http://news.bbc.co.uk/2/hi/science/nature/2804545.stm

RESEARCH ARTICLES

In: Biochemistry and Molecular Biology Coming of Age
Editor: Mary Miu Yee Waye ISBN: 978-1-63321-815-4

Chapter 1

CALPAINS-MEDIATED PTEN CLEAVAGE IN EXCITOTOXIC NEURONAL DEATH

***Minghui Jessica Chen*[#,1], *Piang Chin Benjamin Soo*[#,2], *Meng Shyan Choy*[3], *Zhao Feng Peng*[4], *Heung-Chin Cheng*[2] and *Nam Sang Cheung*[*,1,5]**

[1]Menzies Research Institute School of Medicine, University of Tasmania, Hobart, Tasmania, Australia
[2]Department of Biochemistry and Molecular Biology, Bio21 Institute of Molecular Science and Biotechnology, University of Melbourne, Parkville, Victoria, Australia
[3]Department of Biochemistry, Yong Loo Lin School of Medicine, Faculty of Science, National University of Singapore, Singapore
[4]Key Laboratory of Biogeology and Environmental Geology of the Ministry of Education, China University of Geosciences, Wuhan, P.R. China
[5]School of Life and Environmental Sciences, Deakin University

ABSTRACT

The tumor suppressor PTEN (phosphatase and tensin homolog deleted from chromosome 10) has been implicated in the pathogenesis of numerous neurodegenerative diseases, and is particularly noted for its inhibition of the pro-survival PI3K-Akt pathway. Excitotoxicity due to over-stimulation of ionotropic glutamate receptors in neurons are known to contribute to neurodegenerative diseases such as Alzheimer's disease and ischemic stroke. Previously we have demonstrated that the proteasomal inhibitor, lactacystin, induced cleavage of PTEN to a truncated 50kDa fragment which accumulates to the detergent-insoluble membrane fraction of the neuronal cells. Herein, we report that overstimulation of the ionotropic glutamate receptors with excess NMDA while causing induction of calpains activation in cultured cortical neurons results in the formation of a

[#] Minghui Jessica Chen and Piang Chin Benjamin Soo contributed equally to this study.

[*] Corresponding author and to whom proofs and reprint requests should be addressed: A/Prof Steve Nam Sang Cheung, Menzies Research Institute, School of Medicine Private Bag 23, University of Tasmania, Hobart, Tasmania 7001 Australia. Tel: +613 62262710 Fax: +613 6226 7704 E-mail: nscheung@utas.edu.au.

novel truncated PTEN fragment of molecular weight <50kDa. *In vitro* cleavage assay further confirmed that calpains-mediated cleavage is responsible for the expression of <50kDa, while the 50kDa PTEN fragment observed in latacystin-mediated neuronal death is held accountable by caspase-3 activation. Both truncated PTEN fragments preferentially partition to the detergent-insoluble fraction of neuronal cells. The fragments did not react with the phospho-specific PTEN antibody targeting the C-terminal regulatory serine and threonine residues, suggesting that the cleavage sites are localised at the C-terminal tail or the cleavage enhances dephosphorylation of these C-terminal regulatory sites. Future investigation to define the function and regulation of calpain- and caspase 3-generated truncated PTEN fragment will shed light on the role of PTEN in proteasome inhibition- and excitotoxicity-induced neuronal death in neurological diseases.

Classification Terms: Section, Neurophysiology, Neuropharmacology, other forms of Intercellular Communication

Keywords: PTEN, Truncation, Calpains, Caspase-3

ABBREVIATIONS

Aβ	β-amyloid
α-fodrin	alpha-fodrin
AD	Alzheimer's disease
ASK	apoptosis signal regulating kinase 1
DMSO	dimethyl sulfoxide
ECM	extracellular matrix
ER	endoplasmic reticulum
JNK	c-Jun N-terminal kinase
NB	neurobasal
NMDA	N-methyl-D-aspartate
NP40	Nonidet P-40
PD	Parkinson's disease
PH	pleckstrin-homology
PINK1	PTEN-inducible kinase 1
PIP	phosphatidylinositol-3-phosphate
PIP_2	phosphatidylinositol-3,4-diphosphate
PIP_3	phosphatidylinositol-3,4,5-triphosphate
PI3K	phosphatidylinositol-3-kinase
PTEN	Phosphatase and tensin homologue deleted on chromosome 10
r.t.	room temperature
SDS	sodium dodecyl sulfate
SDS-PAGE	SDS – polyacrylamide gel electrophoresis

INTRODUCTION

Phosphatase and tensin homologue deleted on chromosome 10 (PTEN) is a 403-amino acid multifunctional phospholipid and protein tyrosine phosphatase that is ubiquitously expressed in the neurons (Gary et al., 2002; Luukko et al 1999; Otaegi et al., 2006). PTEN is crucial for the proper neural cerebellar development and architecture, differentiation, cell and organ size regulation (Tamura et al., 1999), and neuronal and glial cell migration (Marino et al., 2002). Its function lies in its ability to specifically hydrolyse the phosphate group from the 3'-OH group of the inositol ring of phosphatidylinositol-3-phosphate (PIP), phosphatidylinositol-3,4-diphosphate (PIP_2) and phosphatidylinositol-3,4,5 triphosphate (PIP_3) through its phospholipid phosphatase activity, with the highest affinity for PIP_3, a major physiological substrate formed by phosphatidylinositol-3-kinase (PI3K) action (Maehama and Dixon, 1998; Myers et al., 1998). PI3K is an important pro-proliferative regulatory molecule that lies upstream of numerous cellular signaling pathways and is involved in regulation of diverse biological functions such as growth regulation, adhesion, motility, invasion, angiogenesis, extracellular matrix attachment and cytoskeletal organization. PIP_3 functions as a potent second messenger that binds and activates a subset of pleckstrin-homology (PH) domain-containing proteins, particularly serine/threonine kinase Akt, to stimulate cellular pathways e.g. mTOR-S6K pathway, that promote oncogenesis (Huang et al., 2005).

PTEN has been implicated in the pathogenesis of numerous neuropathological disorders such as Alzheimer's disease (AD) and ischemic stroke. AD neurons showed a prominent loss of PTEN-immunopositive neurons and altered distribution of PTEN, with subsequent loss of hippocampal CA1 phosphorylated-Akt and PTEN-containing neurons at the end stages of AD (Griffin et al., 2005). These findings are in parallel with observation of increased active phosphorylated Akt in AD in parallel with development of neurofibrillary tangles in a post-mortem brain study (Pei et al., 2003). As such, it can be concluded that aberrant regulation of Akt and PTEN signaling pathway may be one of the events that eventually contribute to manifestation of AD pathology.

On the other hand, in ischemic stroke, PTEN is recently shown to be involved in excitotoxic neuronal death by inhibiting the phosphorylation of serine/threonine kinase Akt (Ning et al., 2004). Gary and colleagues (2002) showed that wild type PTEN predisposed the neurons to cell death when cultured in the presence of glutamate. Neurons that were over expressed with endogenous PTEN displayed lower phospho-Akt compared to neurons that were either untransfected or transfected with dominant negative mutant C147S PTEN. Concurrently, they also showed that in heterozygote PTEN mice; whereby the PTEN expression level is halved compare to homozygote PTEN mice, and exhibited increased of Akt phosphorylation at the serine residue, and showed an increment of neuronal cell viability in the presence of glutamate (Gary et al., 2002). Hence by down-regulating PTEN activities or expression levels the increased resistant to excitotoxic neuronal cell death may be achieved, suggesting that PTEN may be a potential therapeutic target to protect against excitotoxic neuronal death. Further, it has been shown briefly by Zhang and colleagues (2007) that inhibition of calpain 1 by A6185, calpain 1 inhibitor, blocks PTEN degradation (Zhang et al., 2007) in the ischemic brain. However little is known about the role of PTEN in excitotoxic neuronal death (Ning et al., 2004; Choi et al., 2005).

We have previously reported the formation of a 50 kDa truncated PTEN proteolytic fragment upon proteasomal dysfunction by lactacystin which could be abrogated by the pan-specific caspase inhibitor z-VAD-FMK and specific caspase-3 inhibitor IV (Choy et al., 2006). The objective of our study is to elucidate and compare the recruitment of PTEN as a cell death facilitator, in term of protein modifications between two prominent neuronal cell death pathways: namely caspase-3 and calpains which have been widely reported to be activated during neuronal injuries induced upon proteasomal inhibition (lactacystin) and excitotoxic stimulation (N-methyl-D aspartate (NMDA)) respectively. We have demonstrated for the first time, (i) PTEN can be digested by caspase-3 and calpains *in vitro* and in cultured murine primary cortical neurons upon lactacystin and NMDA treatments respectively, resulting in the formation of two novel C-terminal tail-truncated proteolytic fragment of 50 kDa and <50 kDa; (ii) Absence of the C-terminal tail in both novel PTEN species promoting PTEN translocation to the plasma membrane, a finding consistent with that demonstrated in neuronal death mediated by lactacystin.

MATERIALS AND METHOD

2.1. Mouse Neocortical Neuronal Cell Culture Preparation

Neocortical neurons (gestational days 15 or 16) obtained from foetal cortices of Swiss albino mice were used to prepare the primary cultures employing previous described procedures with modifications (Cheung et al., 2000). Microdissected cortices were subjected to trypsin digestion and mechanical trituration. Cells were collected by centrifugation and resuspended in neurobasal™ (NB) medium containing 2.5 % B-27 supplement, 1 % penicillin, 1% streptomycin, 0.25 % GlutaMAX-1 supplement and 10 % dialyzed FCS. 24-well plates prèviously coated with poly-D-lysine (100 μg/ml) were seeded with cells to a density of 2×10^5 cells/cm^2 and used for subsequent experiments. The cultures were maintained in a humidified 5% CO_2 and 95 % air incubator at 37 °C. Immunocytochemical staining of the cultures at day 5 *in vitro* for microtubule-associated protein 2 and glia fibrillary acidic protein revealed >95 % of the cells were neurons with minimal contamination by glia (Cheung et al., 1998). All experiments involving animals were approved by the National University of Singapore, and were in accordance with the US Public Health Service guide for the card and use of laboratory animals.

2.2. Drug Preparation and Treatment

Anhydrous forms of lactacystin (Calbiochem, Cat no. 426100) and NMDA (Sigma-Aldrich, Cat no. M3262) were freshly prepared in dimethyl sulfoxide (DMSO) and water upon usage to achieve a stock concentration of 10 mM and 100 mM respectively. Desired concentrations were achieved through dilution with NB medium. Day 7 neuronal cultures were treated with 1 μM lactacystin and 200 μM NMDA respectively in NB medium for 6 h, 15 h and 24 h.

2.3. Western Blotting

Cells were lysed with RIPA buffer (10 mM Tris HCl, pH 7.4, 1mM EDTA, 150 mM NaCl, 1 % NP40, 0.5 % deoxycholate, 0.1 % sodium dodecyl sulfate (SDS), protease inhibitor cocktail tablet) and spun down at 16000 *g* for 10 min. The detergent-insoluble pellets were re-dissolved in 5X sample buffer (0.5 M Tris (pH 6.8), 10% SDS, 20% glycerol, 0.05 % bromophenol blue, 20 % β-mercaptoethanol) while the concentrations of supernatants were quantitated using Biorad RC-DC assay. After which, the absorbance of each sample solution was read under visible light using the Lowry HS assay. 10 μg of proteins from individual supernatant samples containing 1X SDS (with 20 % freshly added β-mercaptoethanol) and 5 μl of individual pellet samples were heated to 100 °C for 5 min, and allowed to cooled at r.t.. The samples were subsequently centrifuged for 2 min at r.t. at 16000 *g*. The samples were subjected to SDS - polyacrylamide gel electrophoresis (SDS-PAGE; 12 % gel) and immunoblotted using self-raised and commercially available antibodies. The PTEN polyclonal antibody was raised by immunizing rabbits with 5 synthetic peptides from different regions of PTEN. The peptides were K^{352}ANRYFSDNFKVKLYF^{367}, T^{2}AIIKEIVSRNKRRYQED^{19}, T^{232}RREDKFMYFEF^{243}, E^{388}NEPFDEDQHTQITKV^{403} and K^{260}QNKMLKKDKMF^{271}. The phospho-specific anti-phospho-Ser380 PTEN antibody (Cat no. 9551S), and anti-Caspase-3 antibody (Cat no. 9662) were purchased from Cell Signaling. The anti- alph-fodrin (α-fodrin) antibody (Cat no. FG6090) was purchased from BioMol. Anti-tubulin (Cytoskeleton Inc.,, Cat. No. ATN02) monoclonal antibody was used for internal control purpose. All primary antibodies were used at 1:1000 (v/v) dilution in Tris Buffer Saline [0.2 M Tris, 1.5 M NaCl, 0.1 % (v/v) pH 7.4]. Signals were detected using chemiluminescence (Pierce Biotechnology Inc., Rockford, IL, USA) and exposure to X-ray film.

2.4. *In Vitro* Digestion of Recombinant PTEN by Caspase-3, Calpain 1 and Calpain 2

1 μg of recombinant full-length PTEN was incubated with 1 μl of 10X respective protease buffer, 1 μl of protease (caspase-3, calpain 1 or calpain 2; 1 mg/ml) and 4 μl of water for 1h at 23 °C. 5X sample buffer [0.5 M Tris, pH 6.8, 10 % SDS, 20 % glycerol, 0.05 % bromophenol blue, 20 % β-mercaptoethanol] were added to stop the reaction. Proteins in the reaction mixtures were separated by SDS-PAGE, and subjected to Western blotting analysis using anti-PTEN antibody.

RESULTS

3.1. Proteasomal Inhibitor Lactacystin Induced Activation of Caspase 3 and Calapins in Cultured Murine Primary Cortical Neurons

Proteasomal inhibition has been implicated in pathogenesis of neurodegenerative diseases such as AD and Parkinson's disease (PD) (Keller et al., 2000; Moore et al., 2005). Lactacystin is specific, non-reversible and potent proteasome inhibitor that inhibits the chymotrypsin-like and trypsin-like activities of the mammalian 20S proteasome. Dysfunction of the mammalian proteasome function by lactacystin treatment triggers cell death of cultured cortical neurons and cerebellar granule cells via stimulation of mitochondrial cytochrome c release and subsequently caspase-3 protease activity (Pasquini et al., 2000; Qiu et al., 2000). We have previously demonstrated that lactacystin induced a concentration-dependent reduction in cell viability in cultured murine primary cortical neurons and formation of a 50 kDa truncated PTEN proteolytic fragment was observed upon treatment of the neurons with 1 μM lactacystin (Choy et al., 2006). How lactacystin treatment induces truncation of PTEN is unclear. Since PTEN is a key regulator antagonizing the neuronal survival signals emanating from PI3K (Ning et al., 2004) and its down-regulation protects against excitotoxic neuronal death (Ertugrul et al., 2010), we aim to further elucidate the mechanism of lactacystin-induced PTEN truncation in neurons.

There are two major types of proteases contributing to neuronal death – caspase 3 and calpains. In many cell types including neurons and cancer cells, the activated caspase 3 and calpains can proteolyse α-fodrin to give specific cleavage products. Caspase 3 specifically cleaves the Asp^{1185}-Ser^{1186} and Asp^{1178}-Ser^{1179} peptide bonds of α-fodrin, giving rise to 120 kDa and 150 kDa fragments. Calpains, however, cleave α-fodrin at the Tyr^{1176}-Gly^{1177} and Ser^{1231}-A^{1232} peptide bonds to generate only two major fragments of 150 kDa and 145 kDa (Wang, 2000). Thus, the patterns of proteolysis of endogenous α-fodrin in cells have been used by researchers as diagnostic tools to detect activation of caspase 3 and calpains in cells exposed to apoptotic and/or excitotoxic stimuli. For example, the 150, 145 and 120 kDa breakdown products of α-fodrin in cerebrospinal fluid were used as biomarkers of traumatic brain injury associated with aberrant activation of caspase 3 and calpains (Liu et al., 2010).

As shown in Figure 1, lactacystin treatment of primary cortical neurons in cultured neurons for 6 h induced cleavage of endogenous α-fodrin to form three major proteolytic fragments of 150, 145 and 120 kDa.

The abundance of these fragments increased with treatment time. The α-fodrin cleavage pattern suggests that lactacystin treatment induces activation of both caspases and calpains in the cultured neurons. Indeed, the activation of caspase 3 by conversion of pro-caspase 3 to caspase 3 was obvious 15 h after lactacystin treatment through the expression of the 17 kDa active subunit, confirming the caspase 3 as one of the proteases proteolysing α-fodrin in neurons. It is noteworthy that a ~170 kDa fragment of α-fodrin was detectable 6 h after lactacystin treatment. A fragment of similar size was observed when α-fodrin was digested by caspase 3 *in vitro* (Jänicke et al., 1998). The exact identity of this α-fodrin fragment remains unclear. Faint α-fodrin fragments of 170 kDa, 150 kDa, 145 kDa and 120 kDa were noticeable in control cells after 24 h. It is possible that some of the untreated cells undergo apoptosis under the conditions we used, and caspase 3 and calpains were activated in the dying cells. Nevertheless, the truncated α-fodrin fragments were much more abundant in neurons treated with lactacystin.

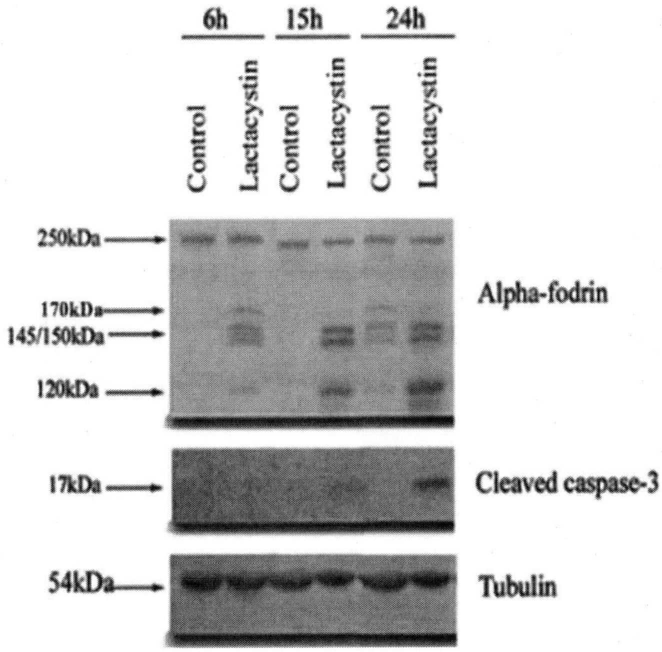

Figure 1. Immunoblotting analysis revealing concurrent involvement of caspase-3 and calpains in proteasomal inhibitor lactacystin-induced neuronal death. Presence of both 120 kDa and 145/150 kDa fragments is indicative of caspase-3 and calpains activation. In addition, the expression of the 17 kDa subunit of active caspase-3 further confirmed its induction in lactacystin-mediated neuronal death. Neurons were treated with 1 μM lactacystin, and incubated at the stated time-points. After which neurons were lysed and used for Western blotting. 10 μg of proteins was loaded per lane. Data were representative of three immunoblots.

3.2. PTEN Is a Common *In Vitro* Substrate to Both Caspase-3 and Calpains - Mediated Cleavage

Previously, we have demonstrated in neurons that during lactacystin-induced proteasomal dysfunction, PTEN is cleaved by caspase-3 –like proteases to form a 50 kDa truncated fragment which could be abrogated by application of caspase inhibitors (Choy et al., 2006). As both caspase-3 and calpains were shown to be activated during lactacystin-mediated neuronal death (Figure 1), in order to ascertain if the shortened 50 kDa PTEN species is a result of caspase-3-induced cleavage, we performed an *in vitro* proteolytic assay on recombinant PTEN employing caspase-3, calpain 1 and 2. As reflected in Figure 2, caspase-3 –generated truncated PTEN protein band (50 kDa) coincided with that being localized in the detergent-insoluble pellet fraction upon lactacystin treatment, confirming caspase-3 as the protease responsible for PTEN truncation in lactacystin-treated neurons. Interestingly, we discovered for the first time, that both calpain 1 and 2 also have the ability to proteolyse recombinant PTEN to generate a truncated PTEN fragment (Figure 2). Furthermore a difference in the molecular size of calpains-generated truncated PTEN fragment (<50 kDa) versus that of caspase-3 (50 kDa) was observed, implying that caspase-3 and calpains cleave PTEN at different sites.

3.3. Calpains-Generated C-Terminal Tail-Truncated PTEN Fragment Is Generated during Excitotoxic Neuronal Injury

To ascertain if calpains-mediated truncated of PTEN is physiologically relevant, we adopted an excitotoxic neuronal model in which calpains but not caspase 3 were activated when the NMDA receptor of the neurons was over-stimulated. NMDA binds and activates NMDA receptors, a major ionotropic glutamate receptor subtype, resulting in increase calcium influx with subsequent activation of calcium-dependent proteases calpains and the calcium-dependent protein phosphatase calcineurin, a process known as excitotoxic cell death.

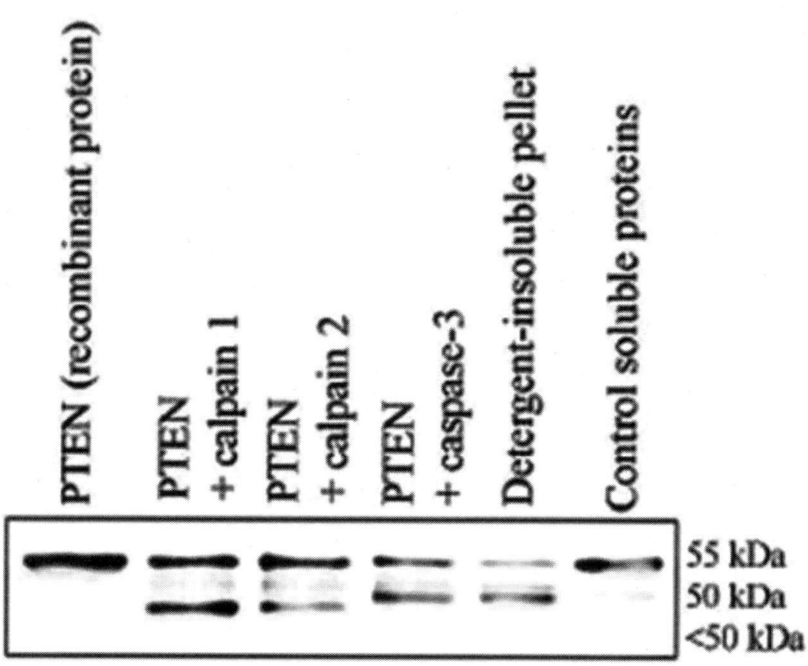

Figure 2. *In vitro* digestion of recombinant PTEN by calpains and caspase-3. Recombinant PTEN (1 µg) was incubated with calpain 1, calpain 2 and caspase-3 for 3 h and the products were analyzed using Western blot. Both calpain 1 and calpain 2 produced truncated PTEN with molecular size of approximately 49 kDa. Caspase-3-digested recombinant PTEN produced 50 kDa truncated PTEN, similar to those observed in the detergent-insoluble pellets of lactacystin-treated cultured cortical neurons. The control soluble proteins were extracted from untreated cultured cortical neurons using RIPA buffer.

Quantitative cell viability assay demonstrated NMDA induced a dose-dependent decrease in survival of cultured murine primary cortical neurons (Carroll et al., 1996). Furthermore, we have previously demonstrated that treatment of cultured primary cortical neurons with 200 µM NMDA induced cellular injury (Carroll et al., 1996). As shown in Figure 3, treatment of cultured cortical neurons with 200 µM of NMDA led to the formation of the 150 kDa and 145 kDa truncated α-fodrin fragments but not the 120 kDa fragments, suggesting that the treatment induced activation of calpains but not caspase 3.

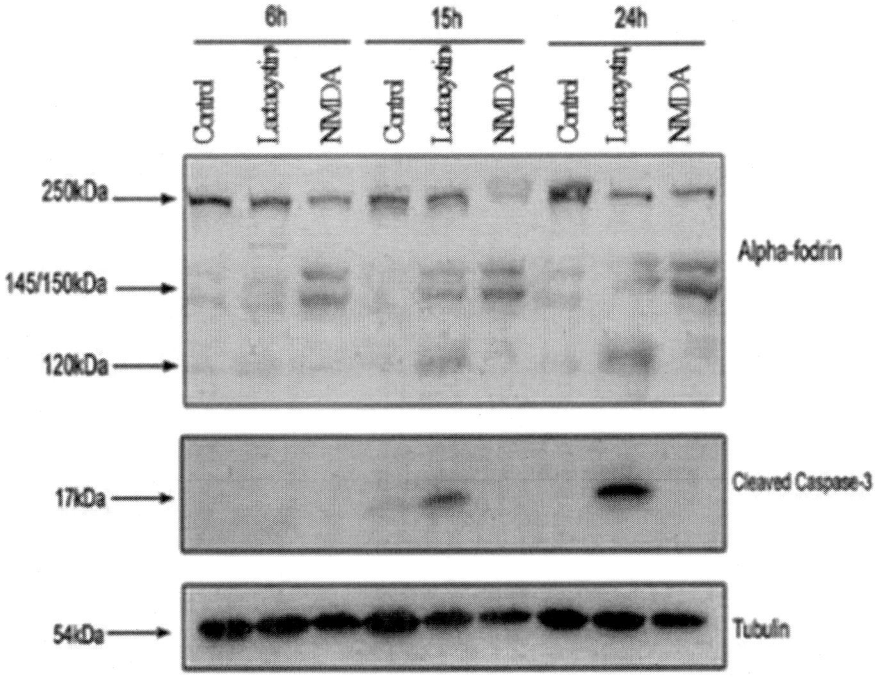

Figure 3. Immunoblotting analysis indicating exclusive calpains activation in NMDA-induced excitotoxic neuronal death. Significant cleavage of α-fodrin to 145/150 kDa fragments is an indication of calpains activation. Absence of the 120 kDa α−fodrin fragment with no active cleaved caspase-3 expression demonstrated NMDA mediation of neuronal death is independent of caspase-3 activation. 250 μM NMDA were applied on cultured primary cortical neurons respectively and harvested at the stated time-points. Lactacystin, a proteasomal inhibitor well established for inducing caspase-3-dependent cell death in neurons was used as a positive control in this blot. 10 μg of proteins was loaded per lane for immunoblotting analysis. Data were representative of three immunoblots.

Previously we reported that the caspase-3 mediated PTEN truncated fragment was specifically recruited to the detergent-insoluble microdomains of the plasma membrane. Presumably, the recruitment permits efficient dephosphorylation of the phospholipid second messengers PIP_2 and PIP_3 in these plasma membrane microdomains. In this study, we aim to determine if the calpain-cleaved PTEN fragment in NMDA receptor-overstimulated neurons also exhibits the same pattern of distribution. Neurons subjected to NMDA and lactacystin time-course treatments were harvested in RIPA buffer containing several detergents (1% Nonidet P-40 (NP40), 0.5% deoxycholate, 0.1% SDS). Cell lysate insolubility in these detergents at 4 °C is attributed to the strong association of cellular proteins with plasma membrane lipid rafts or plasma membrane-associated cytoskeleton. Simultaneous time-course (6 h, 15 h and 24 h) immunoblotting analysis of PTEN expression in the detergent -soluble (comprising mainly of proteins in cytosol and soluble phase of plasma membrane) and – insoluble fractions (proteins in insoluble phase of plasma membrane) of lactacystin and NMDA-treated neuronal sample was carried out (Figure 4). With lactacystin treatment, the truncated PTEN fragment distributed exclusively to the detergent-insoluble microdomains of plasma membrane. Similarly, the truncated PTEN fragment generated by NMDA receptor over-stimulation also distributed to the detergent-insoluble membrane microdomains in neurons.

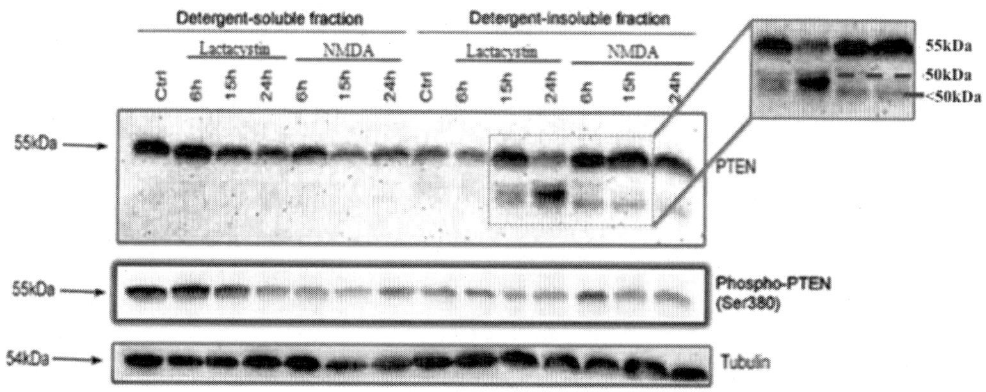

Figure 4. Differential caspase-3 and calpains –mediated PTEN proteolysis at the C-terminal enhanced PTEN distribution from the detergent soluble fraction (represents cytosol and the detergent-soluble microdomains of the plasma membrane) to the detergent-insoluble microdomains of the plasma membrane. Further immunoblotting for phosphorylated PTEN at S380 of the C-terminal tail of PTEN further confirmed the absence of the C-terminal tail in the truncated PTEN species. Neuronal cultures treated with 1 μM lactacystin and 250 μM NMDA were harvested with RIPA buffer over the indicated time-points. The detergent-soluble and insoluble fractions were then subjected to immunoblotting analyses. 10 μg of protein supernatent and 5 ul of the re-dissolved pellet were loaded respectively.

When the samples were analysed by Western blotting using the anti-pS380 phospho-specific antibody that specifically reacts with PTEN when it is phosphorylated at Ser-380 in the C-terminal tail region, the results demonstrate that the antibody was unable to cross-react with the truncated PTEN fragments generated in both treatments. The results suggest that the truncated PTEN fragments were either completely dephosphorylated at Ser-380 or lacking the C-terminal tail region that contains Ser-380.

DISCUSSION

In summary, our study has presented many new findings not previously reported. We have shown that PTEN truncation can be physiologically induced upon caspase-3 and calpains activation, and that the digestion takes place specifically at the C-terminal tail. As the C-terminal tail contains multiple critical phosphorylation sites which govern PTEN migration to the plasma membrane where it exerts its inhibitory effect on the PI3K/Akt pathway, we believe the lack of the C-terminal tail in both novel PTEN species would enhance PTEN translocation to the plasma membrane, resulting in potentiation of neuronal death.

The PI3K/Akt signaling pathway mediates the neuroprotective function of the neurotropic factors. Survival of neurons is maintained by the continuous presence of neurotrophins such as nerve growth factor and brain-derived growth factor (Reichardt, 2006). Studies have shown that PI3K/Akt signaling pathway is important for neurons survival during hypoxic-ischemic brain injury and AD (Delgado-Esteban et al., 2007; Fukunaga and Kawano, 2003; Stein and Johnson, 2002). PTEN enhances neuronal death by down-regulating the PI3K/Akt neuroprotective signaling pathway. PTEN inhibits Akt activities by dephosphorylating PI(3,4,5)P_3 to PI(4,5)P_2, resulting in abolishment the PI3K/Akt pro-survival signals and in turn enhance neuronal death (Delgado-Esteban et al., 2007). To summarize, our unprecedented observations define a new exploratory arena for the study of caspase-3 –PTEN and calpains -PTEN interactions in the promotion of neuronal injury. The implication of these PTEN enzymatic interactions in neuropathology is signified, and can thus be further appreciated by accumulating evidence of the extensiveness of PTEN role in several neurological disorders such as AD, PD and ischemic stroke.

Intra- or extra-cellular accumulation of cross-linked and misfolded proteins is a ubiquitous morphological characteristic of numerous neurodegenerative diseases, including amyotrophic lateral sclerosis, AD, PD and dementia with Lewy bodies (Grune et al., 2004; Sherman and Goldberg, 2001; Shen et al., 2004). This aberrant protein buildup is postulated to be associated with a drop in protein clearance due to inefficiency of the ubiquitin-proteasome system as the use of proteasome inhibitors e.g. lactacystin to suppress proteasome activities in neurons is sufficient reproduce the neuropathological accretion of protein aggregates and subsequently neuronal death (Sawada et al., 2004). PTEN involvement in these protein aggregates-related neurodegenerative diseases is substantiated by the observations of (i) loss and altered distribution of PTEN in AD neurons (Griffith et al., 2005) and (ii) PTEN transcriptional induction of PTEN-inducible kinase 1 (PINK1) which plays a role in neuroprotection against proteasomal dysfunction through inhibition of loss of mitochondrial membrane potential (Shen and Cookson, 2004; Unoki and Nakamura, 2001). Proteasome inhibition has been found to mediate several cell death signaling pathways, including the activation of the cysteine-dependent aspartate-directed (caspase) protease family, the c-jun N-terminal kinase (JNK) pathway, and its upstream kinase and apoptosis signal regulating kinase 1 (ASK) pathway. Recently, studies have also proposed the endoplasmic reticulum (ER) stress-mediated cell death pathway to be involved in the toxicity of proteasome inhibition (Nakagawa et al., 2000; Rao et al., 2005). Caspase-12 is an ER-localized caspase, which is activated by ER stress and can possibly lead to the cleavage of caspase-3 (Rutkowski and Kaufman, 2004). Furthermore, ER stress can also cause the disruption of calcium homeostasis in cells, which in turn activates the calcium-regulated protease calpain and subsequently induces cell death (Rutkowski and Kaufman, 2004). This would have explained the observed concurrent activation of caspase-3 and calpains in lactacystin-mediated neuronal death.

Studies have also shown that over-expression of PTEN enhances excitotoxic neuronal death and down-regulation of the function of endogenous PTEN prevents excitotoxic neuronal death (Gary and Mattson, 2002; Lee et al., 2004; Ning et al., 2004; Zhang et al., 2007). During ischemic insults such as stroke and cardiac arrest, blood flow is disrupted. Cerebral blood flow impairment reduces oxygen and glucose delivery and leads to loss of ATP production in neurons and glial cells. At low ATP concentration, the re-uptake of glutamate into neurons and glial cells is impaired resulting in the accumulation of glutamate in the extracellular space. In addition, deprivation of oxygen and glucose supply can further injure neurons and glial cells. The injured neurons release excessive amount of glutamate to the extracellular space. Taken together, the excessive release of glutamate from the injured neurons as well as impaired re-uptake of glutamate from the extracellular space cause a significant increase in extracellular glutamate concentration to high micromolar or low millimolar range. Excessive glutamate induces over-stimulation of the ionotropic glutamate receptors, especially the abundant NMDA receptor subtype, of neighbouring uninjured neurons which leads to substantial influx of calcium ions from the ECM and intracellular release from organellar stores such as endoplasmic reticulum, resulting in excitotoxicity.

Elevated cytosolic calcium level activates a variety of pro-death components such as calpains and calcineurin. Calpains, a highly conserved family of calcium-activated proteases comprising of two constitutively and ubiquitously expressed calpain isoforms namely, calpain 1 and calpain 2 that are modulated by calpastatin, an endogenous inhibitory protein. Calpains have been implicated in the mediation of the apoptosis in neurons (Crocker et al., 2003) and cleavage endogenous caspases such as caspase-3, -7, -8 and -9 (Chua et al., 2000; McGinnis et al., 1999). Amidst the cleavage of other downstream protein targets, we have shown here that calpains are capable of modulation of PTEN activity and subcellular localization by direct cleavage of the latter's C-terminal tail.

In order to understand the mechanistic implication behind caspase-3 and calpains – generated truncated PTEN fragments, it is important to understand the structure of a full-length PTEN molecule. A schematic diagram of PTEN sequence and its interactive sites is shown in Figure 5. PTEN structural domains include an N-terminal phosphatase domain, a lipid-binding C2 domain, and a 50-amino acids carboxyl-terminal tail that contains a PDZ binding sequence. The 55 kDa cytoplasmic protein shows high resemblance to the dual-specificity phosphatase due to its central catalytic domain exhibiting the signature motif HCXXGXXR in the active sites of protein tyrosine phosphatase. The 190 amino acid region in N-terminal of PTEN containing this signature motif also shares high homology to auxilin and tensin.

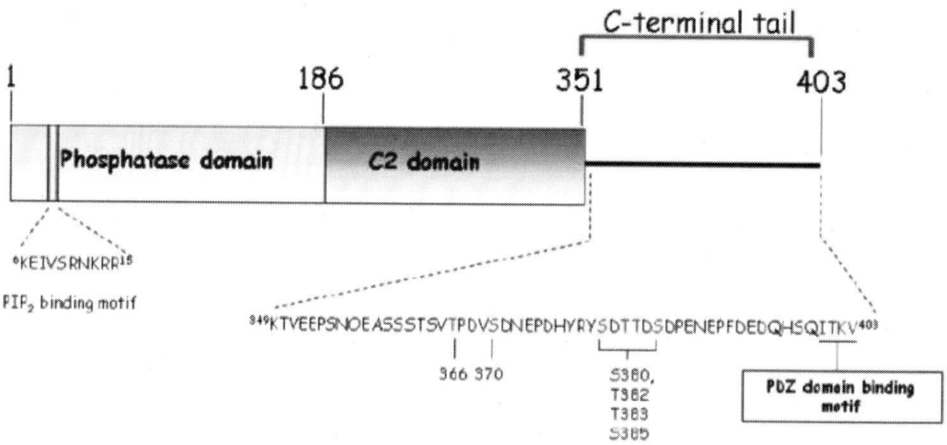

Figure 5. The two structural domains and regulatory motifs of PTEN. The 403 amino acid residue long PTEN is divided into 3 regions: an N-terminal phosphatase domain (residue 1-185), a C2 domain (residues 186-350) and the C-terminal tail (residues 353-403). Residue 6-15 is believed as putative polybasic motif that binds the phospholipid phosphatidylinositol-4,5-bisphosphate on plasma membrane. The C-terminal tail contains several residues, highlighted in red, that undergo phosphorylation and dephosphorylation to regulate PTEN activity and subcellular localisation. The PDZ domain binding motif at the c-terminus, (highlighted in blue) is believed to mediate PTEN binding to unknown PDZ domain-containing proteins.

The C2 domain (residues 186-351) present in PTEN is involved in the mediation of PTEN interaction with the cell membrane together with the phosphatase domain due to the presence of phospholipids-binding motifs responsible for the targeting of PTEN to plasma membrane and its activation. The 50 amino acid PTEN carboxyl-terminal tail (residue 353-403) is implicated in the binding with numerous PDZ domain-containing proteins such as members of the palmitoylated and plasma membrane-associated guanylate kinases comprising of MAGI-2, MAGI-3, and hDLG through the physiological phosphorylation sites Ser-380, Thr-382 and Thr-383 and the ITKV motif, and such interactions may have the ability to regulate PTEN biological activity (Odriozola et al., 2007; Vazquez et al., 2001). Phosphorylation suppresses its activity and prevents its movement to the plasma membrane [Odriozola et al., 2007; Vazquez et al., 2001]. There are six phosphorylation sites in the C-terminal domain that are involved in modulating the activity of PTEN. They include Thr-366, Ser -370, Ser-380, Thr-382, Thr-383 and Ser-385. Studies have shown that substitution of the aforementioned amino acids with alanine (i) abolishes phosphorylation *in vivo*, (ii) enhanced PTEN phosphatase activity, (iii) increases affinity to plasma membrane and (iv): reduces protein stability (Vazquez et al., 2000; Vazquez et al., 2001). Substitution of Ser-380 and Ser-385 with alanine, separately showed an increase in affinity to membrane for PTEN. Substitution of Ser-370, Ser-380, Thr-382 and Thr-383 with alanine separately showed a decrease in PTEN catalytic activity. Besides regulating the activity and affinity to membrane, the C-terminal tail of PTEN also interacts with the C2 domain of PTEN. This intramolecular interaction forms a close monomer complex, masking the PDZ binding domain and thus attenuating PTEN interaction with the PDZ domain containing proteins.

Due to the absence of the C-terminal tail, regulatory phosphorylation to suppress PTEN catalytic activity and eventually inhibit its translocation to the plasma membrane would not be possible. As such, we hypothesized that the proteases-mediated truncated PTEN proteolytic fragment would be more functionally active and exhibit a higher propensity for translocation to the plasma membrane when it exerts its suppression of the PI3K/Akt pro-survival cascade. Since PTEN phosphorylation and dephosphorylation are reversible processes, the regulation of PTEN through phosphorylation might play a role in PTEN tumor suppressor activity. On the other hand, the deletion of the C-terminal tail during neuronal injury is most likely to be associated with cell death since this process is not reversible. Cells undergoing cell demise probably activate proteases such as caspase-3 and calpains to cleave PTEN, causing the latter to translocate to the plasma membrane to antagonize the PI3K/Akt survival pathway.

CONCLUSION

As this is the first report of caspase-3 and calpains -mediated cleavage of PTEN, future investigation should be focus on determining the physiological relevance of these proteolytic fragments in terms of a) its intrinsic activity, b) binding affinity to other interacting partners and c) docking at the plasma membrane, in the aggravation of neuronal injury under neuropathological conditions such as AD, PD and ischemic stroke.

ACKNOWLEDGMENT

This work is supported by Strategic Initiative Funding (Menzies Research Institute) and National Health and Medical Research Council (ID # 509146), Australia.

References

Carroll, F.Y., Beart, P.M., Cheung, N.S. 1996. NMDA-mediated activation of the NO/cGMP pathway: characteristics and regulation in cultured neocortical neurones. *J. Neurosci. Res.* 43, 623-31.

Cheung, N. S., Beart, P. M., Pascoe, C. J., John, C. A., Bernard, O. 2000. Human Bcl-2 protects against AMPA receptor-mediated apoptosis. *J. Neurochem.* 74, 1613-1620.

Choi, J.S., Park, H.J., Kim, H.Y., Kim, S.Y., Lee, J.E., Choi, Y.S., Chun, M.H., Chung, J.W., Lee, M.Y. 2005. Phosphorylation of PTEN and Akt in astrocytes of the rat hippocampus following transient forebrain ischemia. *Cell Tissue Res.* 319, 359-366.

Choy, M.S., Bay, B.H., Cheng, H.C., Cheung, N.S. 2006. PTEN is recruited to specific microdomains of the plasma membrane during lactacystin-induced neuronal apoptosis. *Neurosci. Lett.* 405, 120-125.

Chua, B.T., Guo, K., Li, P. (2000) Direct cleavage by the calcium-activated protease calpain can lead to inactivation of caspases, *J. Biol. Chem.* 275, 5131–5135.

Crocker, S.J., Smith, P.D., Jackson-Lewis, V., Lamba, W.R., Hayley, S.P., Grimm, E., Callaghan, S.M., Slack, R.S., Melloni, E., Przedborski, S., Robertson, G.S., Anisman, H.,

Das, S., Dixon, J. E., Cho, W. 2003. Membrane-binding and activation mechanism of PTEN. *Proc. Natl. Acad. Sci. USA.* 100, 7491-7496.

Delgado-Esteban, M., Martin-Zanca, D., Andres-Martin, L., Almeida, A., Bolanos, J. P. 2007 Inhibition of PTEN by peroxynitrite activates the phosphoinositide-3-kinase/Akt neuroprotective signaling pathway. *J. Neurochem.* 102, 194-205.

Fukunaga, K., Kawano, T. 2003. Akt is a molecular target for signal transduction therapy in brain ischemic insult, *J. Pharmacol* Sci. 92, 317-327.

Gary, D. S., Mattson, M. P. 2002. PTEN regulates Akt kinase activity in hippocampal neurons and increases their sensitivity to glutamate and apoptosis, *Neuromolecular Med.* 2, 261-269.

Griffin, R.J., Moloney, A., Kelliher, M., Johnston, J.A., Ravid, R., Dockery, P., O'Connor, R., O'Neill, C. 2005. Activation of Akt/PKB, increased phosphorylation of Akt substrates and loss and altered distribution of Akt and PTEN are features of Alzheimer's disease pathology. *J. Neurochem.* 93, 105–117.

Grune, T., Jung, T., Merker, K., Davies, K.J.A. 2004. Decreased proteolysis caused by protein aggregates, inclusion bodies, plaques, lipofuscin, ceroid, and 'aggresomes' during oxidative stress, aging, and disease. *Int. J. Biochem. Cell Biol*, 36, 2519 – 2530.

Huang, Y., Wernyj, R.P., Norton, D.D., Precht, P., Seminario, M.C., Wange, R.L. 2005. Modulation of specific protein expression levels by PTEN: identification of AKAP121, DHFR, G3BP, Rap1, and RCC1 as potential targets of PTEN. *Oncogene.* 24, 3819-3829.

Jänicke, R.U., Ng, P., Sprengart, M.L., Porter, A.G. 1998. Caspase-3 is required for alpha-fodrin cleavage but dispensable for cleavage of other death substrates in apoptosis. *J. Biol. Chem.* 273, 15540-15545.

Keller, J.N., Hanni, K.B., Markesbery, W.R. 2000. Impaired proteasome function in Alzheimer's disease, *J. Neurochem.* 75, 436–439.

Kilic, E., ElAli, A., Kilic, U., Guo, Z., Ugur, M., Uslu, U., Bassetti, C.L., Schwab, M.E., Hermann, D.M. 2010. Role of Nogo-A in neuronal survival in the reperfused ischemic brain. *J. Cereb Blood Flow Metab.* 30, 969-984.

Lee, J. H., Kim, K. Y., Lee, Y. K., Park, S. Y., Kim, C. D., Lee, W. S., Rhim, B. Y., Hong, K. W. 2004. Cilostazol prevents focal cerebral ischemic injury by enhancing casein kinase 2 phosphorylation and suppression of phosphatase and tensin homolog deleted from chromosome 10 phosphorylation in rats. *J. Pharmacol. Exp. Ther.* 308, 896-903.

Liu, C., Wu, J., Xu, K., Cai, F., Gu, J., Ma, L., Chen, J. 2010. Neuroprotection by baicalein in ischemic brain injury involves PTEN/AKT pathway. *J. Neurochem.* 112, 1500-1512.

Luukko, K., Ylikorkala, A., Tiainen, M., Makela, T. P. 1999. Expression of LKB1 and PTEN tumor suppressor genes during mouse embryonic development. *Mech. Dev.* 83, 187-190.

Maehama, T., Dixon, J.E. 1998. The tumor suppressor, PTEN/MMAC1, dephosphorylates the lipid second messenger, phosphatidylinositol-3,4,5-trisphosphate. *J. Biol. Chem.* 273, 13375–13378.

Marino, S., Krimpenfort, P., Leung, C., van der Korput, H.A., Trapman, J., Camenisch, I., Berns, A., Brandner, S. 2002. PTEN is essential for cell migration but not for fate determination and tumourigenesis in the cerebellum. Development. 129, 3513-3522.

McGinnis, K.M., Gnegy, M.E., Park, Y.H., Mukerjee, N., Wang, K.K. 1999. Procaspase-3 and poly(ADP)ribose polymerase are calpain substrates. *Biochem. Biophy. Res. Commun.* 263, 94–99.

Merali, Z., Park., D.S. (2003) Inhibition of calpains prevents neuronal and behavioral deficits in an MPTP mouse model of Parkinson's disease, J. *Neurosci.* 23, 4081-4091.

Moore, D.J., West, A.B., Dawson V.L., Dawson, T.M. 2005. Molecular pathophysiology of Parkinson's disease, *Annu. Rev. Neurosci* 28, 57–87.

Myers, M.P., Stolarov, J.P., Eng, C., Li, J., Wang, S.I., Wigler, M.H., Parsons, R., Tonks, N.K. 1997. PTEN, the tumor suppressor from human chromosome 10q23, is a dual-specificity phosphatase. *Proc. Natl. Acad. Sci. USA.* 94, 9052–9057.

Nakagawa, T., Zhu, H., Morishima, N, Li, E., Xu, J., Yankner, B.A., Yuan, J. 2000. Caspase-12 mediates endoplasmic-reticulum-specific apoptosis and cytotoxicity by amyloid-β. *Nature,* 403, 98 – 103.

Ning, K., Pei, L., Liao, M., Liu, B., Zhang, Y., Jiang, W., Mielke, J. G., Li, L., Chen, Y., El-Hayek, Y. H., Fehlings, M. G., Zhang, X., Liu, F., Eubanks, J., Wan, Q. 2004. Dual neuroprotective signaling mediated by downregulating two distinct phosphatase activities of PTEN. *J. Neurosci.* 24, 4052-4060.

Odriozola, L., Singh, G., Hoang, T., Chan, A. M. 2007. Regulation of PTEN activity by its carboxyl-terminal autoinhibitory domain. *J. Biol. Chem.* 282, 23306-23315.

Otaegi, G., Yusta-Boyo, M. J., Vergano-Vera, E., Mendez-Gomez, H. R., Carrera, A. C., Abad, J. L., Gonzalez, M., de la Rosa, E. J., Vicario-Abejon, C., de Pablo, F. 2006. Modulation of the PI 3-kinase-Akt signalling pathway by IGF-I and PTEN regulates the differentiation of neural stem/precursor cells. *J. Cell Sci.* 119, 2739-2748.

Pasquini, L.A., Moreno, M.B., Adamo, A.M., Pasquini, J.M., Soto, E.F. 2000. Lactacystin, a specific inhibitor of the proteasome, induces apoptosis and activates caspase-3 in cultured cerebellar granule cells. *J. Neurosci. Res.* 59, 601–611.

Pei, J.J., Khatoon, S., An, W.L., Nordlinder, M., Tanaka, T., Braak, H., Tsujio, I., Takeda, M,, Alafuzoff, I., Winblad, B., Cowburn, R.F., Grundke-Iqbal, I., Iqbal, K. 2003. Role of protein kinase B in Alzheimer's neurofibrillary pathology. *Acta Neuropathol* (Berl). 105, 381-392.

Qiu, J.H., Asai, A., Chi, S., Saito, N., Hamada, H., Kirino. T. 2000. Proteasome inhibitors induce cytochrome c-caspase-3-like protease-mediated apoptosis in cultured cortical neurons. *J. Neurosci.* 20, 259–265.

Rao, R.V., Ellerby, H.M., Bredesen, D.E. 2004. Coupling endoplasmic reticulum stress to the cell death program. *Cell Death Differ*, 11, 372 – 380.

Reichardt, L. F. 2006. Neurotrophin-regulated signalling pathways. *Philos Trans R Soc Lond B Biol. Sci.* 361, 1545-1564.

Rutkowski, D.T., Kaufman, R.J. 2004. A trip to the ER: coping with stress. *Trends Cell Biol.* 14, 20 – 28.

Sawada, H., Kohno, R., Kihara, T., Izumi, Y., Sakka, N., Ibi, M., Nakanishi, M., Nakamizo, T., Yamakawa, K., Shibasaki, H., Taniguchi, T., Shimohama, S. 2004. Proteasome mediate dopaminergic neuronal degeneration, and its inhibition causes α-synuclein inclusions. *J. Biol. Chem.* 279, 10710 – 10719.

Shen, J., Cookson, M.R. 2004. Mitochondria and dopamine: new insights into recessive parkinsonism. *Neuron.* 43, 301-304.

Shen, X., Zhang, K., Kaufman, R.J. 2004. The unfolded protein response – a stress signaling pathway of the endoplasmic reticulum. *J. Chem. Neuroanat.* 28, 79 – 92.

Sherman, M.Y., Goldberg, A.L. 2001. Cellular Defenses against Unfolded Proteins: A Cell Biologist Thinks about Neurodegenerative Diseases. *Neurons.* 29, 15 – 32.

Sladowski, D., Steer, S. J., Clothier, R. H., Balls, M. 1993. An improved MTT assay, *J. Immunol. Methods.* 157, 203-207.

Stein, T.D., Johnson, J.A. 2002. Lack of neurodegeneration in transgenic mice overexpressing mutant amyloid precursor protein is associated with increased levels of transthyretin and the activation of cell survival pathways. *J. Neurosci.* 22, 7380-7388.

Tamura, M., Gu, J., Takino, T., Yamada, K.M. 1999. Tumor suppressor PTEN inhibition of cell invasion, migration, and growth: differential involvement of focal adhesion kinase and p130Cas. *Cancer Res.* 59, 442-449.

Unoki, M., Nakamura, Y. 2001. Growth-suppressive effects of BPOZ and EGR2, two genes involved in the PTEN signaling pathway. *Oncogene* 20, 4457-4465.

Vazquez, F., Grossman, S. R., Takahashi, Y., Rokas, M. V., Nakamura, N., Sellers, W. R. 2001. Phosphorylation of the PTEN tail acts as an inhibitory switch by preventing its recruitment into a protein complex, *J. Biol. Chem.* 276, 48627-48630.

Vazquez, F., Ramaswamy, S., Nakamura, N., Sellers, W. R. 2000. Phosphorylation of the PTEN tail regulates protein stability and function. *Mol. Cell Biol.* 20, 5010–5018.

Wang, K. K. 2000. Calpain and caspase: can you tell the difference? *Trends Neurosci.* 23, 20-26.

Zhang, Q. G., Wu, D. N., Han, D., Zhang, G. Y. 2007. Critical role of PTEN in the coupling between PI3K/Akt and JNK1/2 signaling in ischemic brain injury. *FEBS Lett.* 581, 495-505.

In: Biochemistry and Molecular Biology Coming of Age
Editor: Mary Miu Yee Waye
ISBN: 978-1-63321-815-4

Chapter 2

NUDCD2 IS OVEREXPRESSED IN HEAD AND NECK SQUAMOUS CELL CARCINOMA AND IS INVOLVED IN NUCLEAR DIVISION[#]

Anne Cromer[1], Alberto Zambrano[1], Régine Millon[2], Joseph Abecassis[2] and Bohdan Wasylyk[1*]

[1]Institut de Génétique et de Biologie Moléculaire et Cellulaire, CNRS/INSERM/ULP, 1 Rue Laurent Fries, Illkirch cedex, France

[2]UPRES EA 34-30, Centre Paul Strauss, 3 rue de la Porte de l'Hôpital, Strasbourg, France

ABSTRACT

Head and Neck squamous cell carcinoma (HNSCC) is the sixth most common cancer in men and has a poor prognosis. The molecular events leading to HNSCC need to be understood. We have identified by Differential Display many novel deregulated genes with unknown functions that require characterization. We show here that one of these genes, *NUDCD2*, is overexpressed in 70% of patients with HNSCC and in many HNSCC cell lines in culture. *NUDCD2* RNA and protein localize to tumour cell islets of human HNSCC compared to surrounding stromal tissue. NUDCD2 overexpression in cells alters cell and nuclear shapes, and induces multinucleation. Time-lapse microscopy showed that there are mitotic aberrations and retention of nucleoplasmic bridges (NPB) between un-separated nuclei following abnormal mitosis. NUDCD2 possesses a p23-like domain, linking NUDCD2 to a family of proteins related to p23, a co-chaperone for Hsp90.

[#] Financial support: (a) the Ligue Nationale Française contre le Cancer, the Ligue Régionale contre le Cancer (Haut-Rhin et Bas-Rhin), the Centre National de la Recherche Scientifique, the Institut National de la Santé et de la Recherche Médicale, the Association pour la Recherche sur le Cancer, Procore Project N° 22010TL; (b) the Ministère de la Recherche et Technologie and the Association pour la Recherche sur le Cancer for fellowships for CA.

[*] Requests for reprints: Bohdan Wasylyk, IGBMC, 1 rue Laurent Fries, BP 10142, 67404 Illkirch, France. Tel: + 33 3 88 65 34 11. Fax: + 33 3 88 65 32 01. Email: boh@igbmc.fr.

Interestingly, the NUDCD2 C-terminal region is also highly similar to the Nuclear Move domain of NudC, a protein involved in the nuclear migration cycle and whose overexpression provokes multinucleation. Furthermore, we found that both proteins colocalize in the Golgi apparatus. These findings suggest that NUDCD2 is a new member of the NudC sub-family that plays an important role in the reformation and maintenance of nuclear integrity following mitosis. Imbalances in the levels of NUDCD2 could be evaluated as a factor contributing to chromosomal instability in HNSCC through effects on the complex processes of mitosis and cell division.

Keywords: HNSCC, overexpression, multinucleation, NudC

Abbreviations

HNSCC	Head and Neck Squamous Cell Carcinoma
RT-PCR	reverse transcription-PCR
RPLP0	Ribosomal Protein Large P0

Introduction

Head and neck squamous cell carcinoma (HNSCC) has an incidence of about 780000 new cases per year in the world, and has a poor rate of survival. HNSCC arises from the surface epithelium of the upper-aerodigestive tract (hypopharynx, pharynx, larynx and oral cavity). Hypopharyngeal HNSCC accounts for about 30% of HNSCC and has a particularly bad prognosis with a high rate of local recurrence and metastasis. There is a need for a better understanding of HNSCC, for the development of rational targeted interventions and for the identification of new prognostic and diagnostic markers [1]. The identification and functional characterization of new genes that are differentially expressed in HNSCC could help in defining the mechanisms of tumourigenesis and tumour progression, and also in the development of targeted therapies. Comprehensive screens of the HNSCC transcriptome using differential display (DD) and DNA microarrays [2-4] have pinpointed a number of genes that are upregulated in this cancer compared with the corresponding normal tissue from the same patients. Amongst them, we selected for further study a novel gene that was recently named *NUDCD2* because it resembles *NudC*.

NudC was initially characterized in Aspergillus nidulans, where it plays a role in nuclear movement following mitosis. Human NudC is overexpressed in proliferating cells, including myeloid and erythroid precursors in normal bone marrow and malignant cells of transformed bone marrow [5]. It is thought that NudC is involved in microtubule and nuclear functions that allow nuclear movement toward daughter cells [6]. Decreasing NudC expression inhibits proliferation and leads to aberrations in mitotic spindle architecture [7], whereas increasing its expression inhibits the growth of prostate tumour cells *in vitro* and *in vivo* in sub-cutaneous xenografts in mice [8]. Increasing or decreasing NudC levels provokes formation of multinucleated cells [9]. Cytokinesis appears to be regulated by phosphorylation of NudC by Plk1 (Polo-like kinase 1) [10], a kinase that is important for maturation and duplication of centrosome, for bipolar spindle formation, for the APC (anaphase promoting complex) complex activation and for sister chromatids segregation [11].

NudC is a member of a new family of proteins that have domains similar to a 150 amino acid sequence of p23, an Hsp90 co-chaperone [12]. Family members have distinct roles, including: yeast kinetochore assembly (Sgt1), β-catenin degradation (Sip), plant pathogenic resistance (Rar1), interaction with integrin β-1 (melusin), and oxydo-reduction (B5+B5R). This dispersion of functions suggests that the p23 domain may mediate protein folding or protein-protein interactions, but its function have not been investigated. The p23-like domain of NudC is also similar to the larger Nuclear Move domain, whose function has not been investigated in mammals.

We show here that NUDCD2 is overexpressed in HNSCC tumours and cell lines. NUDCD2 resembles NudC in that it is localized in the Golgi apparatus, and its overexpression provokes the formation of multinucleated cells, induces aberrant cell division and leads to the accumulation of polyploid cells, perhaps through defective nuclei separation. These results provide clues to further investigating whether NUDCD2 could contribute to aneuploidy and genomic instability, which may provide new avenues for the development of markers and therapeutic strategies.

Materials and Methods

Patient samples. HNSCC tumour samples and the corresponding histologically normal tissues were derived from surgical resections of HNSCC (informed consent was given in all cases). The patients had not been treated at the time of surgery, but were subsequently treated with radiotherapy. The samples were from the advancing edges of the tumours excluding the necrotic centers, and were comprised of 70-80% cancer cells in almost all cases. Normal samples were collected from the farthest margin of the surgical resections (usually uvula).

Differential Display. All tumour samples and corresponding histologically normal tissue (uvula) were obtained, with informed consent. Differential Display was performed using the RNAimage kit (GenHunter Corporation, Nashville, TN) [13]. Amongst the 174 primer combinations used [2], *NUDCD2* was isolated with HT11C (AAGCTTTTTTTTTTTC) and HAP-55 (AAGCTTACGTTAG). Differential bands were cloned into pGEM-T easy (Promega corporation Madison, WI).

Reverse Northerns. Reverse Northern methodology [14] was used to identify which of the 8 clones per DD band are truly differentially expressed. Target sequences were PCR amplified, spotted on nylon membrane and hybridized with radioactive probes prepared from pools of tumour or normal samples. SMART-DD probes were made by: amplifying 200 ng of RNA by SMART (Clontech laboratories, Mountain View, CA); PCR labeling with 3 μl of a HAP primer mix (20 μM), 3 μl of a HT11 primer mix (20 μM), 1 μl of Taq polymerase, 5μl of [^{32}P] dCTP, 1.5 μl of d(G,A,T)TP (20 mM), 3 μl of dCTP (120 μM), 6 μl of 10x PCR buffer, and 1 μl of SMART cDNA (100 ng/μl) using the same cycling steps as in the DD protocol, and purification on Sephadex G50 columns. Membranes were incubated at 65°C for 1 hour and then with probes overnight (buffer: 0.1 g/ml dextran sulfate, 1% SDS, 58 mg/ml NaCl, 0.1 mg/ml salmon sperm DNA). They were washed 3 times for 20 min at 65°C in 2x

SSC, 0.1% SDS, then 3 x 20 min at 25°C with 0.1x SSC, 0.1% SDS and exposed with PhosphoImager screens.

Virtual Northerns. Membranes were prepared as described previously [2, 15]. Briefly, SMART cDNAs from patient samples were run on agarose gels and transferred to nylon membranes (Hybond N+). Hybridization and washes were performed as for "reverse northern" blotting. Probes were generated by random prime labeling (ReadyPrime, Amersham Piscataway, NJ) of DNA containing the *NUDCD2* ORF (Hind III digest) that had been run on agarose gels and purified (Nucleospin Extract, Macherey-Nagel, Easton, PA). Membranes were hybridized and washed as for "reverse northerns".

Quantitative reverse transcriptase PCR (QRT-PCR). Reverse transcription was performed using random primers and the Superscript II RT-PCR system (Invitrogen Life Technologies, Carlsbad, CA), starting from one μg of total RNA. PCR reactions were performed using the Light Cycler system (Roche Diagnostics, Basel, Switzerland). Twenty μl of reaction mixes (4% glycerol; 4mM $MgCl_2$; 200 μM dNTP; 0.15μg/μl BSA; 1x PCR buffer; 0.25x SYBR Green I; 0.025 U of Taq polymerase; 1/200 anti-Taq antibody, 0.5 μM reverse and forward primers) in capillary tubes containing 2 μl of the reverse transcription product diluted 1:50 or 1:250 were subjected to PCR. The PCR conditions were progressive heating from 65 to 95°C, 40 cycles of 10 min at 95°C (1 cycle), 2 sec at 95°C, 10 sec at 60-64°C, 1 sec per 25 base for elongation at 72°C, and cooling. The primers for *NUDCD2* (forward: GAACATGGACTTTGGAGGACA and reverse: TGCTCCACTGAAGTCAAAACC) were chosen with Primer3 software (http://frodo.wi.mit.edu/cgi-bin/primer3/primer3_www.cgi). Expression levels were normalized to the RPLP0 gene (forward: GGAATGTGGGCT TTGTGTTC ; reverse: T GCCCCTGGAGATTTTAGTG).

Immunohistochemistry. Slides containing 0.5 μm paraffin sections of human hypopharyngeal cancer were: deparafinized, rehydrated, pressure cooked for 15 min in citrate buffer pH 6, cooled down for 30 min, incubated for 10 min in 3% H_2O_2, washed 2 x 5 min with Tris-buffered saline pH8.0/0.05% Tween20, incubated for 18 hours at 4°C with primary antibody (rabbit anti-NUDCD2 or preimmune serum) diluted 1:2000 in Dakocytomation Antibody Diluent with background reduction components, washed twice with Tris-buffered saline pH8.0, stained with the LSAB2 (Dako) and DAB (Vector laboratories, Burlingame, CA) kits, counterstained with heamatoxylin, dehydrated, and observed with a 60x objective.

In situ Hybridization. Paraffin sections from human hypopharyngeal cancer (0.5μm) were deparafinized, rehydrated, denatured in 0.2N HCl for 10 min and digested with proteinase K (1μg/ml) for 20 min at 37°C. Digoxigenin labeled RNA probes were prepared with reagents and instructions from Roche Diagnostics by in vitro transcription of the NUDCD2 ORF. The slides were hybridized at 50°C for 18 hours, washed 3x 20 min in 2xSSC/50% formamide at 65°C, rinsed 2x15 min with 1xSSC at 25°C, incubated with an anti-digoxigenin antibody, stained with NBT-BCIP and counterstained with Nuclear Fast Red (Vectors Laboratories, Burlingame, CA).

Cloning of NUDCD2. The 5' extremity of *NUDCD2* was amplified from a human prostate cDNA library (Marathon-Ready TM cDNA, CLONTECH, Mountain View, CA) using the Marathon Adaptor Primer (5' primer [CCATCCTAATACGACTCA CTATAGGGC]) and a *NUDCD2*-specific primer (3' primer [TGAGATTT CTGCTCCACTGAAGTCAAAACC]). The 3' extremity was amplified in the same manner with the Marathon Adaptor Primer and a *NUDCD2* specific primer (5' primer [CTGATGAGGGAACA-TGGACTTTGGAGGACA]). The polymerase chain reaction was carried out using the Advantage cDNA polymerase mix from Clontech. The resulting DNA fragments were cloned into pGEMt-easy. The *NUDCD2* open reading frame was cloned from a head and neck tumour SMART amplified cDNA pool using *NUDCD2* gene-specific primers including BamH1 restriction sites: 5' primer [CGCGGATCCATG-TCGGTCCCGTTTGAG], 3' primer [CGCGGATC-CGTTATTTCTCAAGGTTTG]). The resulting clone in the BamH1 digested pSG5-puromycin vector contained the 471 bp ORF.

Plasmids construction. To add a flag tag to the amino terminus of NUDCD2, a 480 bp fragment of *NUDCD2* was amplified by PCR creating a BamH1 site adjacent to the codon for the first amino-acid of NUDCD2 (5' primer CGCGGATCCATGTCGGTCCCGTTTGAG and 3' primer CGCGGATCCGTTATTTCTCAAGGTTTG). The PCR product was cloned into the pGEMt-easy vector using the pGEMt-easy PCR cloning kit. The BamH1 fragment from pGEMt-easy-*NUDCD2,* containing the ORF region of *NUDCD2,* was ligated into BamH1-digested pSG5-puromycin, pSG5-puromycin-flagNterm, generating *NUDCD2* sense and antisense clones. Plasmids encoding NudC-WT and NudC-mut were provided by T. Zhou (Cambridge, MA). Sequences were subcloned into pSG5-puromycin by inserting a Bam-H1 restriction site by PCR. PCR generated DNA was verified by sequencing.

Generation of anti-NUDCD2 antibody. The peptide CWVQDQMQRKLTLERFQKEN that corresponds to NUDCD2 amino-acid sequence from 113 to 131 was synthesised, coupled to ovalbumin (Imject Maleimide, Pierce Biotechnology, Rockford, IL) and used to immunize rabbits. Western blots of flag-NUDCD2 transfected cells detected 2 bands of around 17 and 19 kDa, as expected for the endogenous 157 amino-acid NUDCD2 protein and 170 amino-acid protein corresponding to the ORF of NUDCD2 fused to flag epitope.

Cell culture and transfection. Cell were cultured at 37°C in a 5% CO2 incubator in Dulbecco's medium supplemented with 2.5% fetal calf serum, 2.5% newborn calf serum and gentamycin (HeLa), or MEM (Invitrogen/GIBCO, Carlsbad, CA) containing 10% fetal calf serum, 1 mM sodium pyruvate, non-essential amino-acids and gentamycin (Hep-2). Transfections were performed using the JetPei reagent according to the manufacturer's recommendations (PolyPlus Transfection, Illkirch, France). Cells were split into 6 well plates 24 hours before transfection. Where indicated, cells were incubating cells for 4 hours at 37°C with 5 μg/ml Nocodazole (Sigma, St. Louis, MO) or 5μg/ml Brefeldin A (Sigma, St. Louis, MO).

Single stable transfectants. Stable Hep-2 transfectants were obtained by puromycin selection of Hep-2 transfected empty pSG5, pSG5-NUDCD2 (sense and antisense), pSG5-flag-NUDCD2 (sense). Ten individual clones and 3 pools were isolated per transfection. 3 individual clones were randomly considered for this study.

Growth curves. Cells were grown at a density of 10^4 cells/wells in 96-well plates. At each time point, cells were treated with 10 μl of CCK-8 (Alexis Biochemicals, San Diego, CA) for 1 hour and the OD was measured at 450 nm using a microplate reader.

Double stable transfectants. Hep-2 cells transfected for 48 h with pBOS-H2BGFP (BD Biosciences, San Jose, CA) [16]. Stably transfected cells were selected with 2 μg/ml of Blasticidin, isolated and submitted to an additional stable transfection. Empty pSG5, pSG5-*NUDCD2* sense and pSG5-*NUDCD2* antisense were transfected for 48 hours and selected with 2 μg/ml puromycin.

Time lapse imaging. Double transfectants grown in 6-well plates to 60-70% confluence were analyzed with an inverted microscope and a 40x objective enclosed in a temperature and carbon-dioxide-controlled incubator. Images were acquired with a CCD camera at intervals of

10 min to 48 hours and analyzed using Metamorph software (Molecular Devices, Downingtown, PA).

Flow cytometry. Cells were cultured in 100 mm plates until subconfluent and washed with 0.1% PBS/EDTA, detached with 1.5 ml of trypsin for 3-5 min at 37°C, and recovered in PBS/10% FCS. 10^6 cells were washed with PBS, pelleted, resuspended in 100 µl PBS and 3 ml of EtOH (70%, -20°C) and incubated for 1 hour at 4°C. Cells were centrifuged (850 rpm), washed with 2% FCS in PBS and stained with a solution containing 100 µl PBS, 100 µl of RNAseA (10 mg/ml) and 200 µl of propidium iodide (100 µg/ml). $2x10^4$ cells were analysed using FacsCalibur and ModFit (BD, San Jose, CA).

Immunofluorescence microscopy. Cells were grown for 24 hours on coverslips, fixed with 3% paraformaldehyde in PBS for 15 min and permeabilized with 0.2% Triton-X100 in PBS for 5 min. Double fluorescence labeling was performed with primary antibodies from rabbit (anti-NUDCD2 or preimmune) or mouse (anti-flag) diluted 1:500 in PBS. The secondary antibodies were either anti-rabbit or anti-mouse coupled to Texas-Red or Fluorescein. Nuclei were stained for 20 sec with Hoechst and Golgi for 30 minutes with WGA-fluorescein (Invitrogen Molecular probes, Carlsbad, CA). No cross-reaction was observed in control experiments. The coverslips were mounted on slides and viewed with a microscope with a 100x objective, filters specific for red (Texas-Red) and green (Fluorescein) fluorophores, and a monochrome camera. The digitalized images were cropped, colored, assembled, and labeled using Adobe Photoshop 6.0.

Quantification of multinucleation. Plasmids containing NUDCD2-sense, NUDCD2-antisense, NudC-flag, no insert and EGFP (co-transfection marker) were transfected into HeLa cells. After 12, 24, 48, 72 and 96 hours the cells were fixed and stained with Hoechst (nuclear marker). The slides were observed by fluorescence microscopy with a 60x objective. Two slides per point were required, because of EGFP extinction. The proportion of cells containing more than 2 nuclei or nuclear abnormalities (multinucleation, micronuclei or enlarged nucleus) were analyzed. A series of 50-150 pictures was acquired, in order to consider at least 200 cells for each plasmid transfection. The 3 images were merged and cell proportions were counted manually.

RESULTS

Isolation and sequence analysis of NUDCD2. We have reported the identification of genes differentially expressed in hypopharyngeal carcinoma by Differential Display technology [2] and by oligonucleotides microarrays [17], and shown that the DD approach has the advantage of identifying novel and uncharacterized genes [4]. For that reason, we focused on one clone (1655E), which was isolated by Differential Display, which was not present on Affymetrix HG-U95A microarrays, and whose nucleotide sequence was not in the databases at the time of identification.

Searches in the Genbank data base did not yield any 1655E homologues, but several human expressed sequence tags were discovered with a high degree of similarity to 1655E, suggesting that the sequence is expressed and might correspond to a new transcript. We used the Differential Display sequence to design 1655E-specific oligonucleotides and used the RACE (rapid amplification of cDNA ends) procedure with human prostate cDNA to amplify

5' and 3' sequences. The resulting cDNA fragments were sequenced and assembled (now corresponding to Genbank AAH17934). Several ESTs had 100% homology to this sequence and large fragments of cDNA sequence were identical to Genbank (NT_023133) genomic sequences, indicating that the gene has 4 exons of 276, 49, 152 and 475 bp, distributed on 13.6 kb of chromosome 5. The cDNA was found to contain an open reading frame encoding a 157-amino-acid polypeptide with a calculated molecular mass of 17.7 kDa (Supplementary Figure 1). Sequence alignment of several species identified by Blast in protein databases revealed that the protein sequence of NUDCD2 is evolutionarily highly conserved (Tables 1 and 2, and Supplementary Figure 1A).

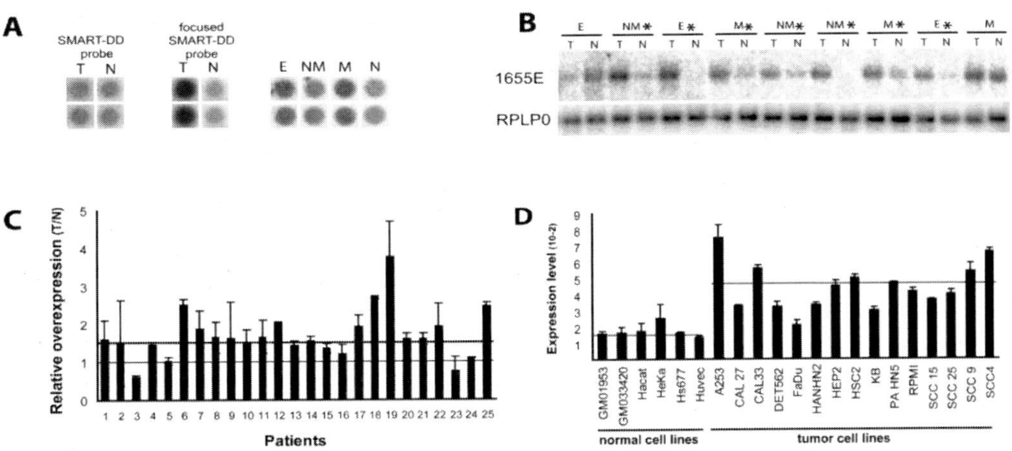

Figure 1. NUDCD2 is overexpressed in HNSCC. *A,* NUDCD2 messenger RNA levels were assessed by Reverse Northern blotting. Duplicate PCR products from the Differential Display were spotted on membranes and hybridized with pools of different samples: T (tumour), N (normal), E (Early), NM (Not Metastasis prone) or M (Metastasis prone). *B,* Virtual Northern blot. T and N represent tumour and normal samples, respectively, from the same patient. RPLP0 was used as a loading control. Patients exhibiting differential expression of NUDCD2 are highlighted with an asterisk. *C, D* Quantitative RT PCR of 25 patient samples (*C*) and cell lines (*D)* for NUDCD2 and RPLP0 for normalization. The data is expressed as mean values with error bars representing standard deviations for two replicate experiments.

Table 1.

Orthologs

Accession number		Species	Similarity (%)	Identity (%)
BI 341915	pig	Sus scrofa	100	99
BAB28205	mouse	Mus musculus	100	99
CB4384444	cow	Bos taurus	100	98
CB 784716	rat	Rattus norvegicus	99	99
BU325397	chicken	Gallus gallus	94	87
AL873013	frog	Silurana tropicalis	91	77
AW642264	frog	Xenopus leavis	89	76
BM183007	zebrafish	Danio rerio	83	70
BX303628	trout	Onchorhynchus mykiss	83	69

Table 2.

Paralogs

Accession number		p23-domain containing protein	Similarity (%)	Identity (%)
Q9H0N2	NudC	involved in cytokinesis	50	27
Q9BVR5	CML66	tumor antigen	48	23
Q9BTI3		hypothetical protein	42	22
Q15185	p23	unactive progesterone receptor, hsp90 co-chaperone	32	17
Q9NQA7	B-ind 1	butyrate-induced 1	31	19
Q9Y2Z0	SGT1	suppressor of the G2 allele of Skp1	31	15
O60066	SIP	calcyclin binding protein	29	13
Q9UKP3	melusin	integrin b1 binding protein 2	28	14
Q9BTE6		hypothetical protein	25	9
NM_130810	EKN1	dislexia susceptibility	24	13
Q9UHI9	B5+B5R	flavohemoprotein	23	16
Q9UHD1	Chp-1	CHORD-containing protein 1	22	13
O94966	USP19	deubiquitinating enzyme 19	21	9

The polypeptide had no predicted transmembrane or other membrane anchoring sequences but was found to contain a putative p23-like domain (cd00237; amino-acid residues 14-104, Supplementary Figure 1B). The gene is now officially named NUDCD2 because its p23-like domain overlaps with a Nuclear Move domain (pfam03593; amino-acid residues 162-315, Supplementary Figure 1C). The presence of this domain defines a new sub-family of p23-like proteins including NudC, NUDCD1, NUDCD2 and NUDCD3 [18], which predictably could also have specific conserved functions.

NUDCD2 is overexpressed in HNSCC solid tumours and cell lines. The differential expression of NUDCD2 was not described in our initial report [2] due to the sensitivity of the probe. We discovered it with an improved methodology, which focuses the labelling to a subset of transcripts by using the same primers and conditions that were used for the Differential Display [4]. NUDCD2 showed increased expression in tumour compared to normal uvula samples (Figure 1A; T, tumour; N, normal). Virtual Northern blot analysis revealed a transcript of about 1 kilobase in length that is overexpressed in most of the tumour samples examined (7/9) compared to matching normal samples from the same patients, except for patients 1 and 9 that showed decreased or equal expression in tumour compared to normal uvula (Figure 1B). Some tumour grade specific expression was initially suspected by Reverse Northern (E=*e*arly, M = *m*etastasis within 3 years after treatment, NM = no metastasis during a minimum 3 year follow-up; Figure 1A), which was not substantiated with the individual patient samples (Figure 1B).

Additional patient samples were analysed by quantitative QRT-PCR. Most of the patients analyzed (17/25) displayed at least 1.5 overexpression in tumours ($p<0.001$, Figure 1C). In addition, NUDCD2 was expressed in tumour cell lines at a higher level (2.6 fold) than in cell lines derived from normal tissue ($p<0.001$, Figure 1D; Supplementary Figure 2). The highest mRNA levels were observed in the A253 salivary gland and SCC4 HNSCC cell lines (Figure 1D). Therefore, we conclude that NUDCD2 is frequently overexpressed in head and neck malignant cells in vivo and in vitro.

NUDCD2 mRNA was detected in HNSCC tumour cells by in situ hybridization (Figure 2A), compared to the sense probe used as a negative control (Figure 2B). The protein was localized by immunohistochemistry with a NUDCD2-specific rabbit polyclonal antibodies that we developed against the carboxy-terminal part of NUDCD2 (amino acids 113 to 131).

The protein was found to be overexpressed in the cytoplasm of malignant cells that form tumour islets and have a dysplastic morphology (Figure 2C). Overexpression was obvious in 5/9 patient samples, slight in 2/9 and not-detectable in 2/9. At this stage we had identified a new gene overexpressed in tumour cells from the hypopharynx that required further functional analysis.

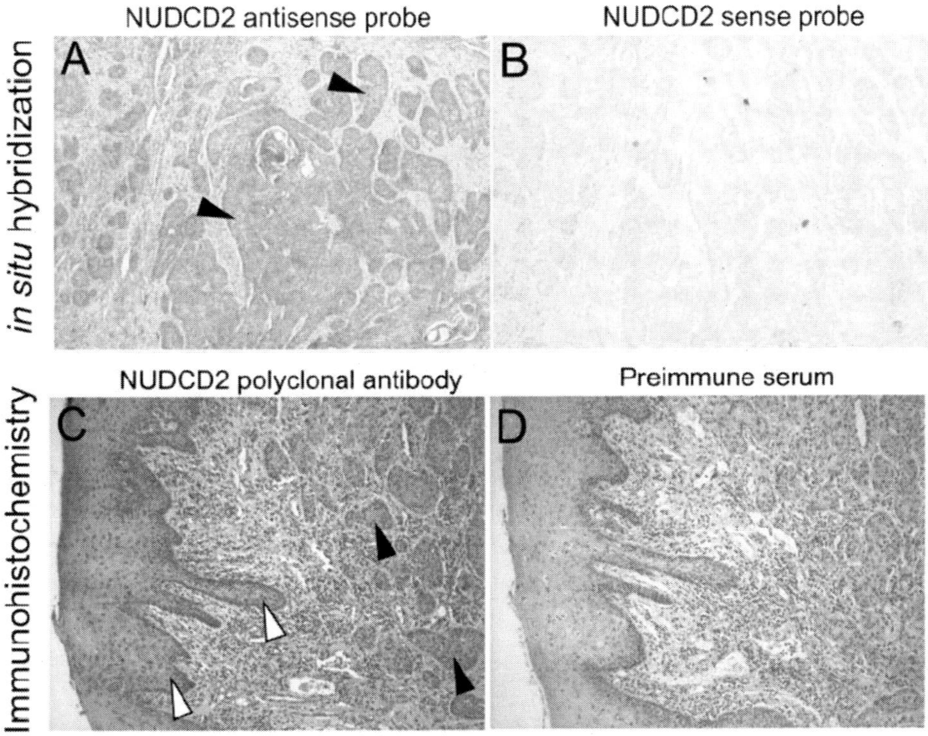

Figure 2. NUDCD2 overexpression in the cytoplasm of hypopharyngeal tumour cells. Tumour sections were analysed by *in situ* hybridisation with antisense *(A)* and sense *(B)* NUDCD2 mRNA probes (blue/purple) and counterstained with nuclear fast red (red/pink). Tumour sections were also analysed by immunohistochemistry *(C, D)* with rabbit anti-NUDCD2 *(C)* or preimmune *(D)* serum (brown stain) and heamatoxylin counterstain (blue/purple). Black arrowheads indicate tumour cells islets and white arrowheads displasia.

Endogenous NUDCD2 is localised in the Golgi apparatus in cells. In order to determine whether NUDCD2 is, like its homologue NudC, a Golgi resident protein, we determined the intracellular localization of NUDCD2 using NUDCD2 antibodies. Indirect immuno-fluorescence confocal microscopy revealed that NUDCD2 is primarily localized in the juxtanuclear region together with wheat germ agglutinin (WGA), a Golgi marker (Figure 3). To more precisely assess the co-localization of NUDCD2 and WGA, we performed shift experiments with pharmacological reagents. Incubation of cells with Nocodazole, which alters microtubule dynamic and disassembles Golgi [19, 20] redistributed NUDCD2 in concert with WGA. Similarly, treatment with Brefeldin A, which disrupts the structure and function of the Golgi [21], resulted in relocalisation of NUDCD2 and WGA to more diffuse structures. Thus, endogenous NUDCD2 is localized to the same Golgi compartment as WGA.

We also studied the localisation of overexpressed NUDCD2 compared to NudC. Transfected flagged NUDCD2 gave a homogenous cytoplasmic distribution with some increased staining in the perinuclear area (data not shown and Figure 4A lower panel). This perinuclear staining could represent the endogenous protein in the Golgi, and cytoplasmic staining to spill over from saturation of the Golgi. It is probably not due to the flag epitope since overexpressed NUDCD2 without the flag displayed similar staining detected (using a NUDCD2 antibody, data not shown).

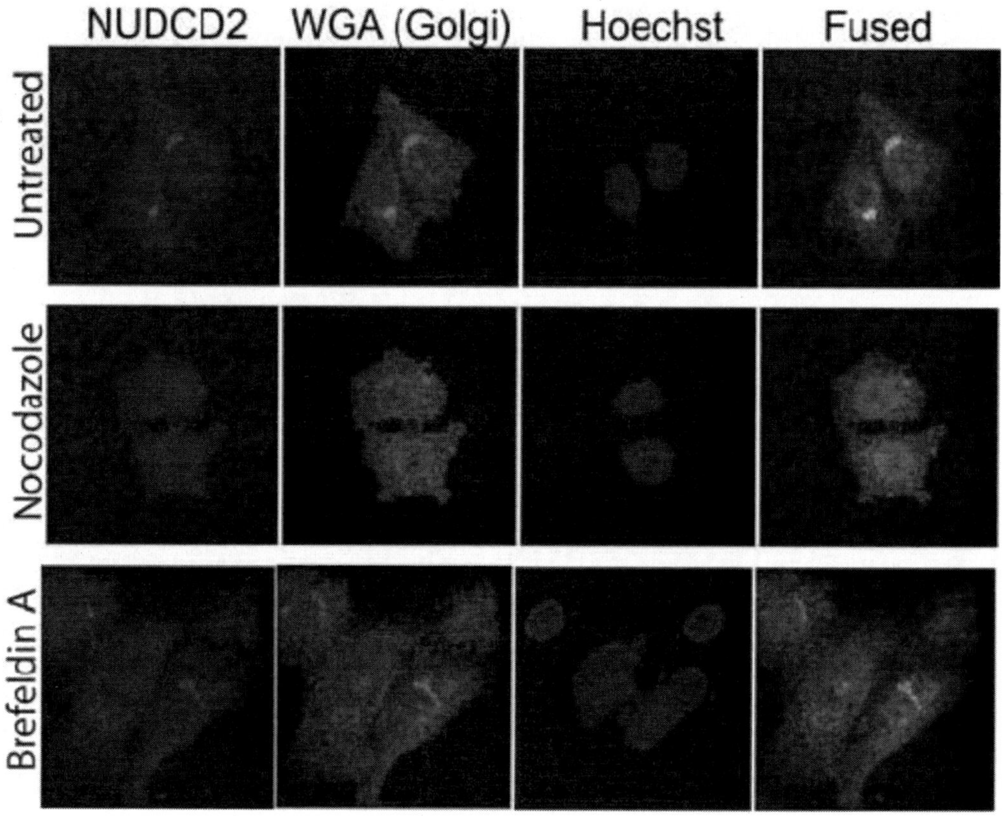

Figure 3. Immunocytochemistry showing that endogenous NUDCD2 is localized in the Golgi apparatus of Hela cells. Stains: NUDCD2 (2133 antibody, Texas Red), Golgi (Wheat germ-agglutinin FITC conjugated, green), nuclei (Hoechst, blue). Separate and fused images are shown. Treatments to disrupt the Golgi: Nocodazole (5μg/ml; 4 h), Brefeldin A (5μg/ml; 4 h).

Rat NudC protein has been reported to reside in the Golgi [22] and human NudC protein with MTOCs in neurons [23]. We found that overexpressed exogenous NudC was homogenously distributed in the cytoplasm and colocalised with NUDCD2 (Figure 4A, lower panel). Contrary to NUDCD2, NudC seems to be excluded from nucleus.

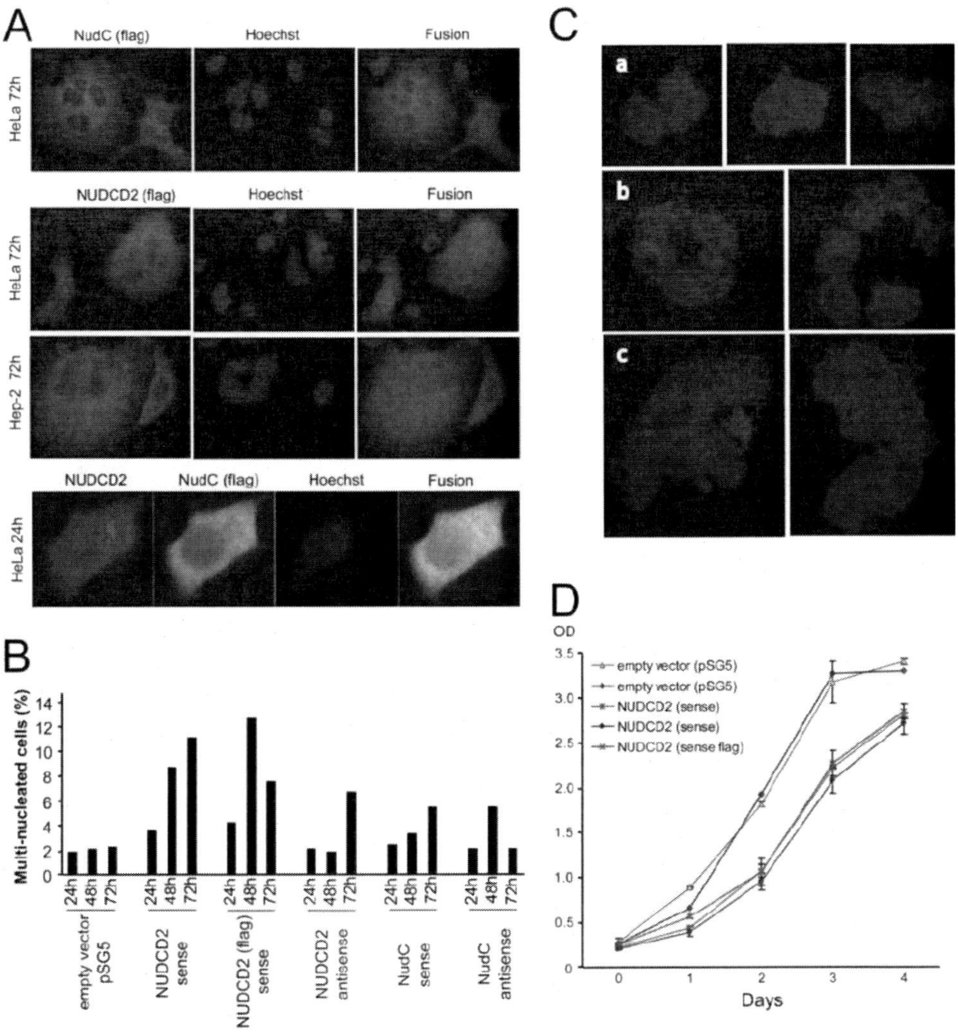

Figure 4. Transiently (*A, B*) or stably (*C, D*) transfected cells were used to show that NUDCD2 is localised in the cytoplam (*A*), induces multinucleation (*A-C*) and slows cell growth (*D*). *A*, NudC and NUDCD2 were expressed in HeLa or Hep-2 cells and after 72 h they were detected using monoclonal antibodies against the flag tag or rabbit antibodies against NUDCD2 and fluorescence labelled secondary antibodies, and observed by microscopy (bottom panels: confocal). *B*, Cells were transfected with NUDCD2 or NudC sense, antisense or empty vectors, and cells bearing multinucleic equivalent to two or more 2 nuclei were quantified at 24 h, 48 h and 72 hours post-transfection. At least 200 EGFP co-transfected cells were counted. *C*, Representative nuclear morphology of stably transfected cells. Up to 50% of overexpressing cells had "butterfly shaped" binucleated nuclei (a, upper panels), around 2% had "ring shaped" mutinucleated cells (b, middle panels) and 8-12% of cells had strongly enlarged mis-shaped nuclei (c, lower panels). Nucleus content was stained using Hoechst (blue). *D*, Stably transfected cells had impaired cell growth compared to controls. Viable cells were measured using the Cell Counting Kit-8.

Overexpression of NUDCD2 provokes multinucleation. Increasing or decreasing NudC expression in HeLa cells has been shown to generate flat and enlarged cells with aberrant and distended multinuclei [9]. We studied the effects of altering NUDCD2 expression by transient transfection, using co-expressed EGFP to identify transfected cells (Figures 4A and B and

Supplementary Figure 3). Increasing or decreasing NUDCD2 levels, like NudC, led to significant increases in the number of cells with aberrant multinuclei and spread enlarged cytoplasms. NUDCD2 was at least as efficient as NudC at inducing these changes. Similar effects were observed in two other cell lines, Hep-2 and FaDu (Figure 4A and data not shown). The proportion of cells with aberrant nuclei was lower than reported by Aumais et al. [9], which may reflect the different methods used to introduce NudC (transient transfection as opposed to viral infection).

The effects of overexpression were also studied by stable transfection in Hep-2, a HeLa subline. We found that the NUDCD2-expressing stable clones had significant and often striking alterations compared to the control clones with empty vector. The clones that emerged during the isolation process were examined morphologically. A significant proportion (15%) were giant cells, with enlarged nuclei and cytoplasms (Figure 4Dc). The most frequent phenotype (60%) was joined bi-nuclei (butterfly nuclei) (Figure 4Da), apparently corresponding to two nuclei that were very close or fused with NPB (nucleoplasmic bridges), or a nucleus with an additional aberrant nuclear peripheral structure. There were also clones with multiple "doughnut" nuclei displayed as a ring (Figure 4Db). These results show that, similar to its homologue NudC, NUDCD2 overexpression provokes the loss of normal nuclear structure.

The isolated clones were examined for effects on cell growth. Clones overexpressing NUDCD2 grew more slowly than control clones containing empty vector (Figure 4C), showing that NUDCD2 also affects cell division.

We performed time-lapse microscopy to establish whether the "giant" cells are syncytia arising from cells fusion or multinucleated cells resulting from aberrant division. Hep-2 cells stably transfected with a istone H2B-GFP fusion protein and either empty vector or NUDCD2 were analyzed by time-lapse microscopy. GFP fluorescence was used to follow the chromatin of mitotic NUDCD2 overexpressing cells. (Figure 5A; cell boundaries were detected by bright-field microscopy and are indicated by white lines). Control clones expressing H2B-GFP alone or H2B-GFP and empty vector were composed almost exclusively of normal-looking cells (about 90 %), with some unusual cells that looked apoptotic or had aberrant mitotic spindles (Other, 9 %; Figure 5C). Normal cells took about two hours to divide (Figure 5Aa). The NUDCD2 overexpressing clones contained a significant proportion of unusual looking cells with joined nuclei or NPB (30 %; Figures 5Ab and c, 5B and 5C). About 60 % of the cells looked normal, and amongst the other cells a small constant proportion (about 10%) appeared to be apoptotic or aberrant (other, 10%, Figure 5C). The formation of joined nuclei or NPB was followed over time (Figures 5Ab and 5Ac). Figure 5Ab illustrates a cell whose nucleus did not separate and became a joined bi-nucleated (butterfly) cell. Figure 5Ac illustrates a tri-nucleated cell that attempted to divide but ended up with a severe multinucleated phenotype. The nuclei did not separate during the time they were observed, which was about 10 times longer than a normal cell division (Figure 5Ab, 2510 min; Figure 5Ac, 2280 min). The nuclei remained attached by bridges (see NBP on enlargements of Figures 5Ab and 5Ac in 5Bb and 5Bc respectively). These results suggest that the multi-nucleation in NUDCD2 overexpressing cells results from a defect in nuclear separation.

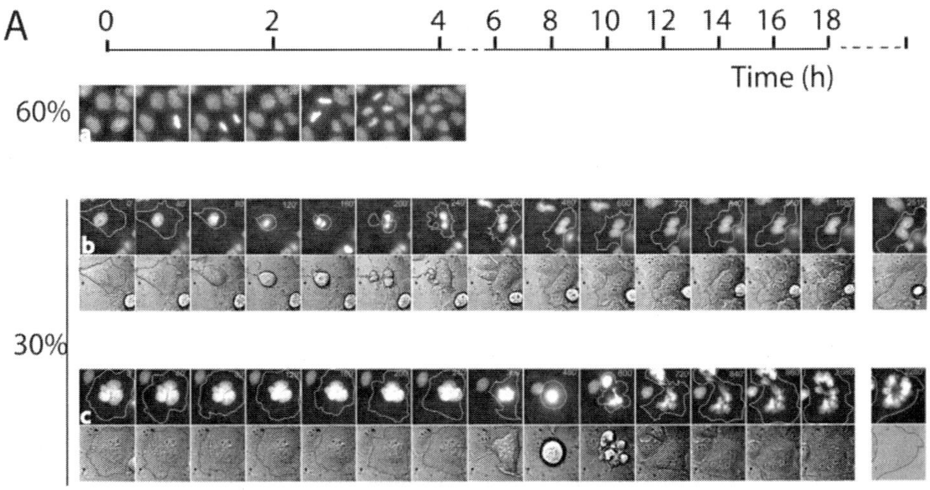

B

H2B-Histone	+	+	+
empty vector	-	+	-
NUDCD2 sense	-	-	+
Normal	**90.8**	**91.3**	**60.2**
Joined nuclei	**0.0**	**0.0**	**30.1**
Other	**9.2**	**8.7**	**9.7**

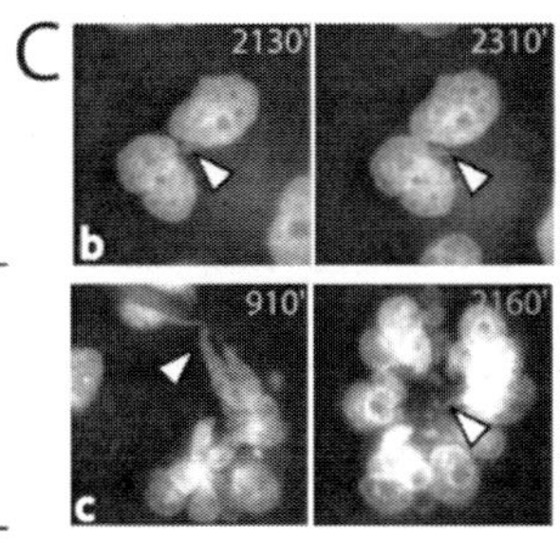

D

H2B-Histone	+	+
NUDCD2 sense	-	+
G1	**49.0** (ND)	**45.4** (7.17)
S	**37.1** (ND)	**36.4** (4.09)
G2/M	**14.1** (ND)	**18.5** (5.33)

Figure 5. Overexpression of NUDCD2 delays mitosis, inhibits nuclear separation and induces multinucleation. Time lapse microscopy of doubly NUDCD2 and Histone-H2B transfected cell lines. *A*, Fluorescent micrographs for GFP labelled histone H2B (upper panels) and phase contrast light micrographs (lower panels) are shown (the cell periphery is indicated by white or black lines, respectively). Representative mitoses of NUDCD2 overexpressing cells are shown over an 18 hour period. *B*, Proportions (%) of cells transfected with the indicated vectors that appear normal, have joined nuclei, or display "other" aberrant morphologies are shown. At least 50 cells were counted. *C*, Enlargements of the late telophase cells shown in *Ab (Cb)* and Ac *(Cc)*Arrowheads shows nuclear bridges which link nearly separated nuclei. *D*, Cell cycle analysis by flow cytometry. Numbers in parentheses are the range of two values.

CONCLUSION

Our previous studies allowed us to uncover several genes whose expression is deregulated in cancer, including many that are novel [4, 24, 25]. The characterization of NUDCD2 revealed that it is a novel gene overexpressed in HNSCC. We showed that it is significantly overexpressed in hypopharyngeal tumours, including "early" tumours (small-sized, T1/T2, moderately to well differentiated, without lymph node involvement), which may indicate an early role in HNSCC tumourigenesis.

A

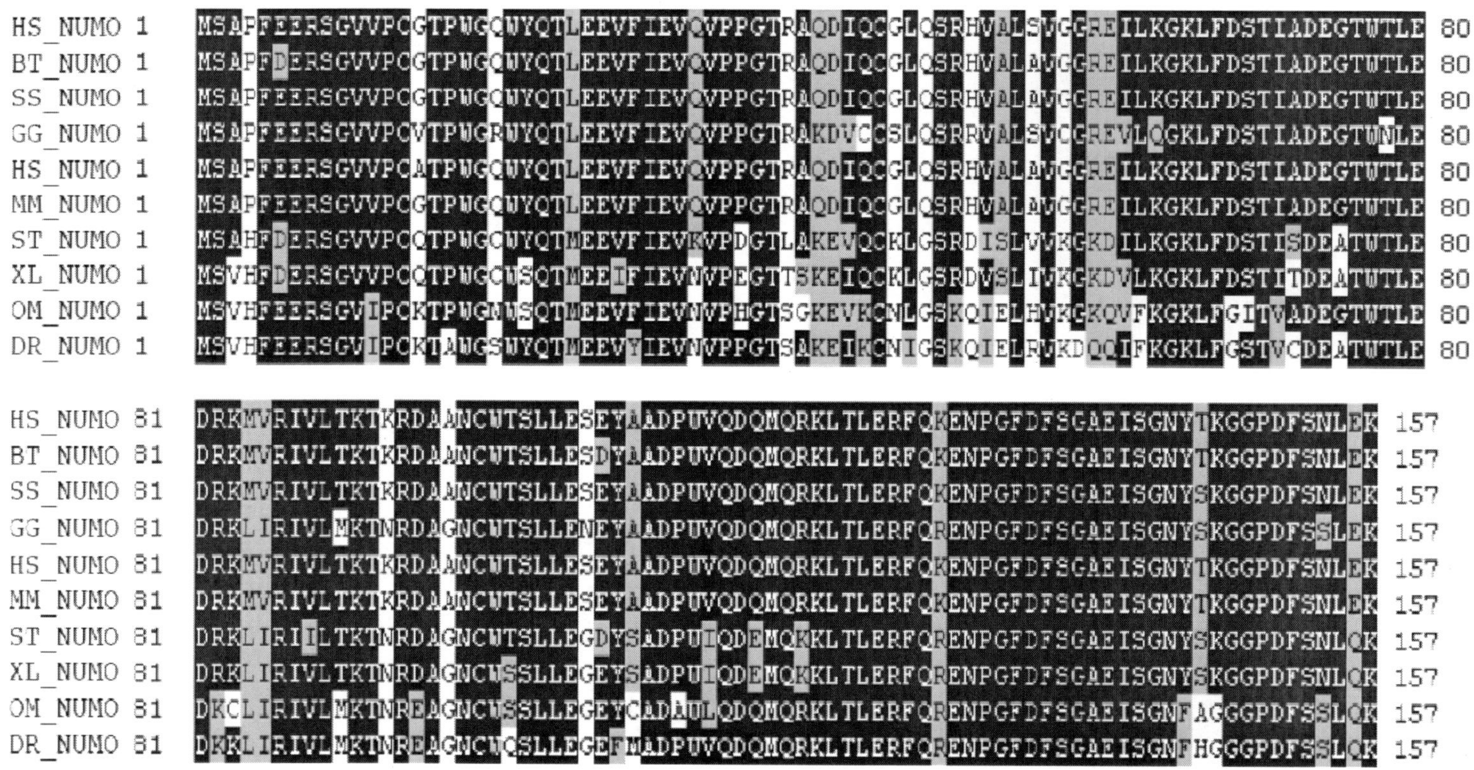

Supplementary Figure 1A & B. NUDCD2 orthologs and paralogs. A, Main NUDCD2 orthologs (HS: Homo Sapiens, BT: Bos Taurus, SS, Sus Scrofa; GG, Gallus Gallus; RN, Rattus Norvegicus; MM, Mus Musculus; ST, Silurana Tropicalis; XL, Xenopus Laevis; OM, Onchorhynchus Mykiss, DR, Danio Rerio). Proteins or EST translated accession numbers are given in Table 1. B, p23 domain of the NUDCD2 paralogs. Accession numbers and name (or symbol) are indicated on the left. Secondary structure ($\beta 1$ to $\beta 7$) of p23 and amino-acids that are conserved, from the 3D structure of p23 and HSP20 [12] are indicated in red (hydrophobic core) and blue (polar amino-acid) in the lower part.

B

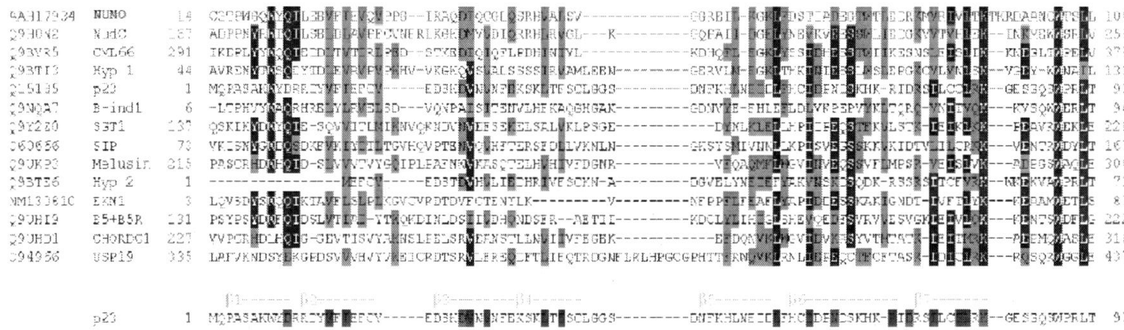

C

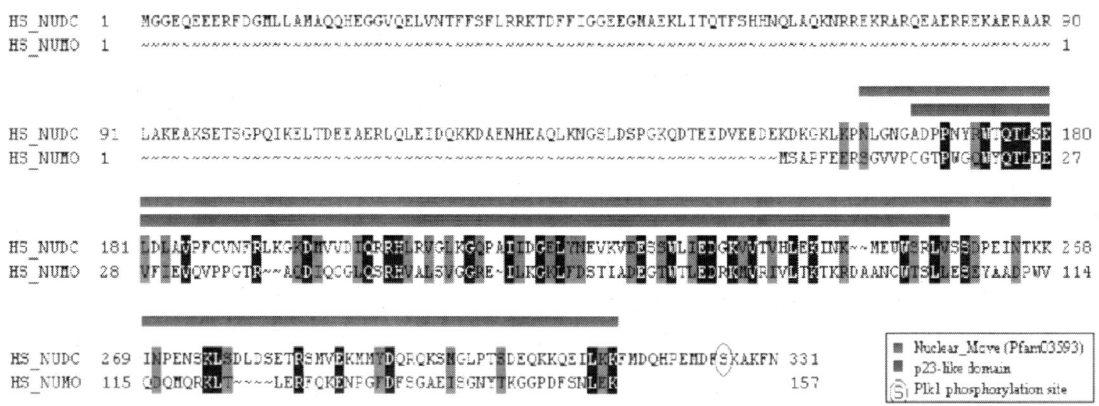

Supplementary Figure 1C. NUDCD2 orthologs and paralogs. C, NUDCD2 and NudC protein sequence alignment. The Nuclear Move domain (Pfam 03593; aa 162-315) covers most of the length of NUDCD2, whereas the p23 domain (cd00237; aa 14-104) is located in the N-terminal part of the Nuclear Move domain. NudC possesses an additional N-terminal part compared to NUDCD2. Plk1 phosphorylation sites of NudC involved in cytokinesis [10] are indicated. Identical and similar amino acids are highlighted in black and gray, respectively.

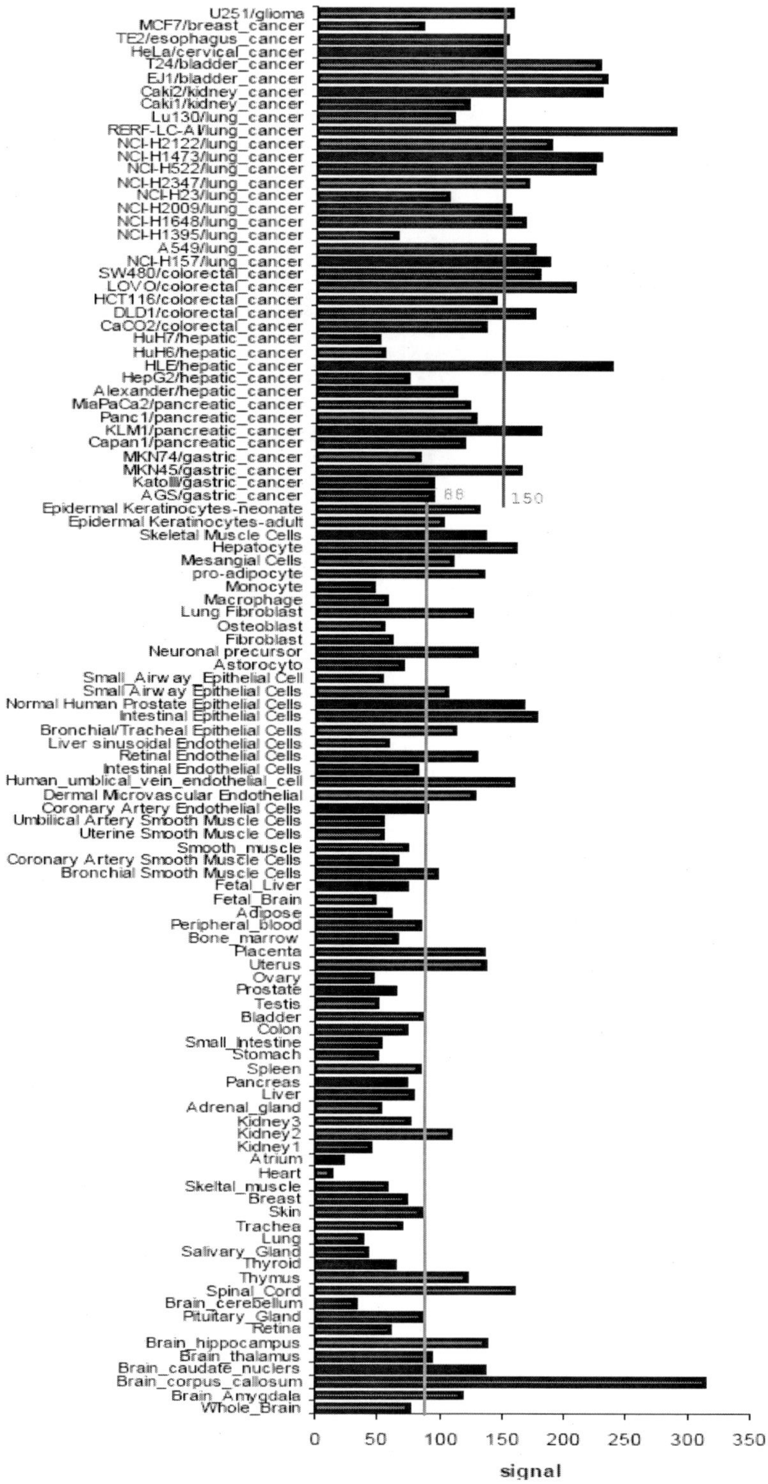

Supplementary Figure 2. NUDCD2 expression in cancer and normal tissues based on public microarray data. Signal represents NUDCD2 expression level in various tissues. Raw data from [36] was available at http://www.lsbm.org. The green bar indicates the average expression value in normal tissues (88) and the red bar indicates the average expression value in cancer cells (150).

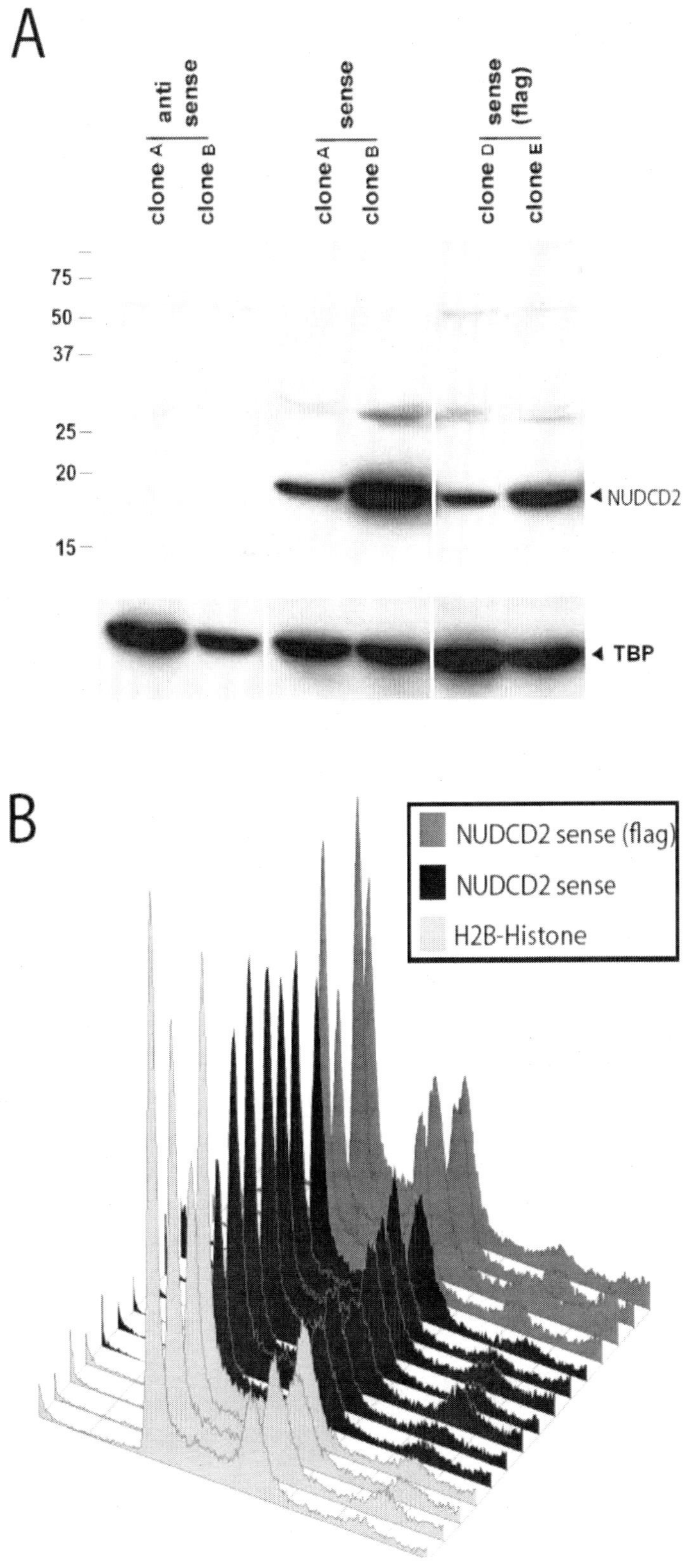

Supplementary Figure 3. (Continued).

C

H2B-Histone	+	+	+
NUDCD2 antisense	-	+	-
NUDCD2 sense	-	-	+
G1	**48.9**	**52.7**	**45.3**
S	**37.1**	**34.2**	**36.4**
G2/M	**14.0**	**12.6**	**18.5**

Supplementary Figure 3. NUDCD2 overexpression increases the proportion of cells in the G2-M phase of the cell cycle. A. Western blot using anti-NUDCD2 antibody showing the average overexpression level of sense-NUDCD2 in representative induividual stable Hep-2 clones. The level of TBP is shown as an internal loading control. B, C. Cell cycle analysis by flow cytometry. The figure shows histograms (B) and population percentage of cells (C) in the different phases of the cell cycle. Ten individual stable Hep-2 clones overexpressing either NUDCD2 sense, antisense or empty vector were analyzed. Defects of nuclear separation should affect the proportion of cells in the different stages of the cell cycle. We analysed vector control and NUDCD2-transfected cells for DNA content analysis by flow cytometry. NUDCD2 overexpressing cells had a significantly higher proportions ($p=0.035$) of cells in the G2-M phase of the cell cycle (4n DNA content), and the DNA content per cell cycle was remarkably polymorphic, suggesting that defective mitosis might have led to polyploidy.

Our time-lapse experiments indicate that NUDCD2 inhibits cell division, apparently upstream of NudC. NUDCD2 overexpression appears to inhibit nuclear separation, whereas NUDC overexpression inhibits cytoplasmic separation, a later step in cell division. NudC is expressed in the midbody during cytokinesis and phosphorylation of NudC by Plk1 is involved in this process [10]. We found that NUDCD2 is not precisely localized to the midbody during cytokinesis (data not shown), and NUDCD2 does not have a sequence related to the NudC Plk1phosphorylation sites (Supplementary Figure 1C). NUDCD2 could be involved in tumourigenesis through its effects on mitosis at the level of cell or nuclear division. Nuclear envelope and nuclear structure reassembly has to proceed in a tightly regulated manner after chromatid separation, ensuring that the interphase organization of chromatin can be re-established in daughter nuclei.

It has been suggested that the activation of DNA-damage response pathways serves as a surveillance mechanism to eliminate cells with tumourigenic potential and that defects in these pathways allow escape of tumourigenic cells. Our observations, together with the fact that the presence of NPB has been validated as a biomarker of DNA damage in human B lymphoblastoid cell line [26] and the observation that NUDCD2 expression is increased in p53 mutant cells (data not shown, [27]), suggest that NUDCD2 overexpression could results in activation of DNA-damage response pathways [28] [29] but it does not show that NUDCD2 could be sufficient to induce DNA damage and the aneuploidy phenotype. This has to be further investigated.

Gisselsson et al. reported that abnormal nuclear morphology (associated with nucleoplasmic bridges, micronuclei, and nuclear buds) is indicative of significant genomic instability within cells and is a common feature of a wide variety of cancers [30]. Defects of nuclear separation should affect the proportion of cells in the different stages of the cell cycle.

We analysed vector control and NUDCD2-transfected cells for DNA content analysis by flow cytometry. NUDCD2 overexpressing cells had a significantly higher proportion ($p=0.035$) of cells in the G2-M phase of the cell cycle (4n DNA content), and the DNA content per cell cycle was remarkably polymorphic (Supplementary Figure 3), suggesting that defective mitosis might have led to polyploidy. Although many cells with defective mitoses will be eliminated by apoptosis or subsequent cell-cycle arrest, some NUDCD2 overexpressing cells with genomic defects may escape these checkpoints, allowing them to proliferate and to form tumours [29]. This scenario is consistent with our observations indicating that increase of NUDCD2 does not lead to a block in cell cycle but merely to a slowed progression through mitosis. One might then predict that any defects that lead to slippage of cells with genomic defects through mitosis can give rise to tumour cells [31].

NUDCD2 displays homologies with the p23 and related proteins from other species including chicken, frog and plants, suggesting that NUDCD2 is a member of a large family of proteins with conserved functions. Human p23 was first identified as a protein associated with the progesterone receptor and was later shown to be an Hsp90 co-chaperone that also possesses a chaperone activity of its own [32]. NUDCD2 could be a chaperone for proteins involved in cell division following mitosis. Chaperones appear to have an essential guardian function that is subverted during oncogenesis to allow malignant transformation and to facilitate rapid somatic evolution. NUDCD2 may be involved in tumourigenesis through these mechanisms. Interestingly, pharmacologically inhibitors of chaperones seem to offer a promising and unique anticancer strategy [33].

The precise role of NUDCD2 in cancer development remains to elucidate, but our data open new directions to study mechanisms involved in late telophase as well as in cancer development. The participation of this protein in nuclear division is an important although perhaps not unique aspect of its activity; NUDCD2 and related proteins may have other cellular functions that remain to be discovered. Interestingly, a recent study has reported that NUDCD2 (called NUDCL2 in this report) appears to be a regulator of the LIS1/dynein pathway in mammals [34]. However, the physiological functions of NUDCD2 were not examined. Our result open up the possibility that regulation of nuclear movement by Lis1 and Dynein [35] could be modulated by NUDCD2, and overexpression of NUDCD2 could perturb this function in cancer.

Our results point to the possibility that NUDCD2 overexpression in HNSCC could contribute to tumourigenesis through its striking effects on the fidelity of nuclear division and chromosome ploidy. They add a new component to the understanding of the molecular processes involved in cancer.

References

[1] Hasina R, Lingen MW. Head and neck cancer: the pursuit of molecular diagnostic markers. *Semin. Oncol.* 2004 Dec;31(6):718-5.

[2] Lemaire F, Millon R, Young J, Cromer A, Wasylyk C, Schultz I, et al. Differential expression profiling of head and neck squamous cell carcinoma (HNSCC). *Br. J. Cancer.* 2003 Nov 17;89(10):1940-9.

[3] Cromer A, Carles A, Millon R, Ganguli G, Chalmel F, Lemaire F, et al. Identification of genes associated with tumorigenesis and metastatic potential of hypopharyngeal cancer by microarray analysis. *Oncogene.* 2004 Apr 1;23(14):2484-8.

[4] Carles A, Millon R, Cromer A, Ganguli G, Lemaire F, Young J, et al. Head and neck squamous cell carcinoma transcriptome analysis by comprehensive validated differential display. *Oncogene.* 2005 Oct 31.

[5] Miller BA, Zhang MY, Gocke CD, De Souza C, Osmani AH, Lynch C, et al. A homolog of the fungal nuclear migration gene nudC is involved in normal and malignant human hematopoiesis. *Exp. Hematol.* 1999 Apr;27(4):742-0.

[6] Osmani AH, Osmani SA, Morris NR. The molecular cloning and identification of a gene product specifically required for nuclear movement in Aspergillus nidulans. *J. Cell Biol.* 1990 Aug;111(2):543-1.

[7] Zhang MY, Huang NN, Clawson GA, Osmani SA, Pan W, Xin P, et al. Involvement of the fungal nuclear migration gene nudC human homolog in cell proliferation and mitotic spindle formation. *Exp. Cell Res.* 2002 Feb 1;273(1):73-4.

[8] Lin SH, Nishino M, Luo W, Aumais JP, Galfione M, Kuang J, et al. Inhibition of prostate tumor growth by overexpression of NudC, a microtubule motor-associated protein. *Oncogene.* 2004 Apr 1;23(14):2499-6.

[9] Aumais JP, Williams SN, Luo W, Nishino M, Caldwell KA, Caldwell GA, et al. Role for NudC, a dynein-associated nuclear movement protein, in mitosis and cytokinesis. *J. Cell Sci.* 2003 May 15;116(Pt 10):1991-03.

[10] Zhou T, Aumais JP, Liu X, Yu-Lee LY, Erikson RL. A role for Plk1 phosphorylation of NudC in cytokinesis. *Dev. Cell.* 2003 Jul;5(1):127-8.

[11] Nigg EA. Polo-like kinases: positive regulators of cell division from start to finish. Curr *Opin. Cell Biol.* 1998 Dec;10(6):776-3.

[12] Garcia-Ranea JA, Mirey G, Camonis J, Valencia A. p23 and HSP20/alpha-crystallin proteins define a conserved sequence domain present in other eukaryotic protein families. *FEBS Lett.* 2002 Oct 9;529(2-3):162-7.

[13] Liang P, Pardee AB. Differential display of eukaryotic messenger RNA by means of the polymerase chain reaction. *Science.* 1992 Aug 14;257(5072):967-1.

[14] Poirier GM, Erlander MG. Postdifferential display: parallel processing of candidates using small amounts of RNA. *Methods.* 1998 Dec;16(4):444-2.

[15] Franz O, Bruchhaus II, Roeder T. Verification of differential gene transcription using virtual northern blotting. *Nucleic Acids Res.* 1999 Jun 1;27(11):e3.

[16] Kanda T, Sullivan KF, Wahl GM. Histone-GFP fusion protein enables sensitive analysis of chromosome dynamics in living mammalian cells. *Curr Biol.* 1998 Mar 26;8(7):377-5.

[17] Cromer A, Carles A, Millon R, Ganguli G, Chalmel F, Lemaire F, et al. Identification of genes associated with tumorigenesis and metastatic potential of hypopharyngeal cancer by microarray analysis. *Oncogene.* 2004 Apr 1;23(14):2484-98.

[18] Zhou T, Zimmerman W, Liu X, Erikson RL. A mammalian NudC-like protein essential for dynein stability and cell viability. *Proc. Natl. Acad. Sci. USA.* 2006 Jun 13;103(24):9039-44.

[19] Vasquez RJ, Howell B, Yvon AM, Wadsworth P, Cassimeris L. Nanomolar concentrations of nocodazole alter microtubule dynamic instability in vivo and in vitro. *Mol. Biol. Cell.* 1997 Jun;8(6):973-5.

[20] Storrie B, Yang W. Dynamics of the interphase mammalian Golgi complex as revealed through drugs producing reversible Golgi disassembly. *Biochim. Biophys Acta.* 1998 Aug 14;1404(1-2):127-7.

[21] Dinter A, Berger EG. Golgi-disturbing agents. *Histochem Cell Biol.* 1998 May-Jun;109(5-6):571-0.

[22] Morris SM, Yu-Lee LY. Expression of RNUDC, a potential nuclear movement protein, in mammalian cells: localization to the Golgi apparatus. *Exp. Cell Res.* 1998 Jan 10;238(1):23-2.

[23] Aumais JP, Tunstead JR, McNeil RS, Schaar BT, McConnell SK, Lin SH, et al. NudC associates with Lis1 and the dynein motor at the leading pole of neurons. *J. Neurosci.* 2001 Dec 15;21(24):RC187.

[24] Lemaire F, Millon R, Muller D, Rabouel Y, Bracco L, Abecassis J, et al. Loss of HOP tumour suppressor expression in head and neck squamous cell carcinoma. *Br. J. Cancer.* 2004 Jul 19;91(2):258-1.

[25] Seghatoleslam A, Zambrano A, Millon R, Ganguli G, Argentini M, Cromer A, et al. Analysis of a novel human gene, LOC92912, over-expressed in hypopharyngeal tumours. *Biochem. Biophys Res. Commun.* 2006 Jan 6;339(1):422-9.

[26] Thomas P, Umegaki K, Fenech M. Nucleoplasmic bridges are a sensitive measure of chromosome rearrangement in the cytokinesis-block micronucleus assay. *Mutagenesis.* 2003 Mar;18(2):187-94.

[27] Robinson M, Jiang P, Cui J, Li J, Wang Y, Swaroop M, et al. Global genechip profiling to identify genes responsive to p53-induced growth arrest and apoptosis in human lung carcinoma cells. *Cancer Biol. Ther.* 2003 Jul-Aug;2(4):406-15.

[28] Gorgoulis VG, Vassiliou LV, Karakaidos P, Zacharatos P, Kotsinas A, Liloglou T, et al. Activation of the DNA damage checkpoint and genomic instability in human precancerous lesions. *Nature.* 2005 Apr 14;434(7035):907-13.

[29] Bartkova J, Horejsi Z, Koed K, Kramer A, Tort F, Zieger K, et al. DNA damage response as a candidate anti-cancer barrier in early human tumorigenesis. *Nature.* 2005 Apr 14;434(7035):864-70.

[30] Gisselsson D, Bjork J, Hoglund M, Mertens F, Dal Cin P, Akerman M, et al. Abnormal nuclear shape in solid tumors reflects mitotic instability. *Am. J. Pathol.* 2001 Jan;158(1):199-206.

[31] Weaver BA, Cleveland DW. Decoding the links between mitosis, cancer, and chemotherapy: The mitotic checkpoint, adaptation, and cell death. *Cancer Cell.* 2005 Jul;8(1):7-12.

[32] Bose S, Weikl T, Bugl H, Buchner J. Chaperone function of Hsp90-associated proteins. *Science.* 1996 Dec 6;274(5293):1715-7.

[33] Whitesell L, Lindquist SL. HSP90 and the chaperoning of cancer. *Nat. Rev Cancer.* 2005 Oct;5(10):761-72.

[34] Yang Y, Yan X, Cai Y, Lu Y, Si J, Zhou T. NudC-like protein 2 regulates the LIS1/dynein pathway by stabilizing LIS1 with Hsp90. *Proc. Natl. Acad. Sci. USA.* 2010 Feb 23;107(8):3499-504.

[35] Lee WL, Oberle JR, Cooper JA. The role of the lissencephaly protein Pac1 during nuclear migration in budding yeast. *J. Cell Biol.* 2003 Feb 3;160(3):355-64.
[36] Ge X, Yamamoto S, Tsutsumi S, Midorikawa Y, Ihara S, Wang SM, et al. Interpreting expression profiles of cancers by genome-wide survey of breadth of expression in normal tissues. *Genomics.* 2005 Aug;86(2):127-41.

In: Biochemistry and Molecular Biology Coming of Age
Editor: Mary Miu Yee Waye ISBN: 978-1-63321-815-4

Chapter 3

GENOTYPE ANALYSES USING SNP (USING MALDI-TOF MASS SPECTROMETRY) AND STR (MICROSATELLITE) MARKERS IN THE DETERMINATION OF ZYGOSITY STATUS OF CHINESE TWINS

Cadmon K. B. Lim[1], Venus S. Y. Yeung[1], Tze Lun Yeung[1], Aaron C. Y. Tam[1], Connie SH Ho[2], Simpson W. L. Wong[2], Bonnie W. Y. Chow[2], Yee Miao Ho[2], Crystal Hin Ning Chou[2], Chi-Hang Fred Fung[1], Chee Fei Wong[1] and Mary M. Y. Waye*[*,1]

[1]Croucher Laboratory for Human Genomics, School of Biomedical Sciences, The Chinese University of Hong Kong, Shatin, N.T. Hong Kong
[2]Department of Psychiatry, The University of Hong Kong, Hong Kong

ABSTRACT

It has been argued that single nucleotide polymorphisms (SNPs) is a very useful tool complementary to microsatellite analyses in the near future due to its relative ease of automation and interpretation. However, because allele frequencies can vary greatly among populations, it is important to validate and identify the informative SNPs that are useful for sample identification in biomedical and epidemiological studies. We have genotyped 768 individuals (384 pairs of twins) with a panel of SNPs based mostly on the SNPs chosen by an earlier work conducted by Lee et al., 2005 for Koreans, which we hypothesized to be useful for our Chinese subjects. The MALDI-TOF mass spectrometry based iPLEX Gold assay on the MassARRAY® Platform method using Sequenom was used for determination of the zygosity status of the twins. The ABI microsatellite kit (AmpFISTR identifiler PCR amplifcation kit) for measuring 16 loci was used to confirm

[*] Corresponding author: ary M.Y. Waye, Ph.D. Director, Croucher Laboratory for Human Genomics, Rm 608 MMW Bldg., School of Biomedical Sciences, The Chinese University of Hong Kong, Shatin, N.T. Hong Kong; Tel: (852)2609-6874; Fax: (852)2603-7732; Email: mary-waye@cuhk.edu.hk.

the zygosity status. We found that some of the SNPs (2/22) used in Lee's paper were not suitable for our Hong Kong Chinese samples. The zygosity status as determined by the mass spectrometry method can be validated with microsatellite method using 76 samples. In summary, we have studied a panel of 25 SNPs on their effectiveness in distinguishing identical and non-identical twins (768 individuals) for the Hong Kong Chinese population using the iPLEX method of Sequenom. It is important to optimize SNP panels for genotyping depending on the population and the platform used for the studies.

Abbreviations

SNPs	single nucleotide polymorphisms
STRs	short tandem repeats
STMS	sequence tagged microsatellites sites
DZ	dizygotic
MZ	monozygotic
LD	Linkage disequilibrium
P-E	paternity exclusion
P-I	probability of identity

Introduction

Most scientists agree that single nucleotide polymorphisms (SNPs) are useful tools complementary to current microsatellite technology for forensic studies, and a few even argue that it is likely to gain an increasing prominent role as a fundamental technology in identification of individuals due to its relative ease of automation and interpretation. Even though microsatellites markers has formed the basis of many national databases for forensic purpose, there are some advantages of using SNPs under many experimental situations. Some of the differences of SNPs compared with sequence tagged microsatellites sites (STMS) or short tandem repeats (STRs) have been discussed by Bhat, 2001 and Glover et al. 2010, as follows:

a) SNPs are more stable than STMS or STRs which are more prone to size homoplasy (Angers et al., 2000) and has variation due to mutational patterns, hence SNPs are more reliable.
b) SNPs are much more prevalent than STMS or STRs (Kwok et al., 1996).
c) Frequency of occurrence of SNPs is far higher than that of STMS or STRs (every few tens of kbps) Gibb et al., 2003 and Frazer et al., 2007.
d) SNPs are more non-uniform in distribution. High density in STMS or STRs region in Drosophila (Colson and Goldstein, 1999).
e) SNP density within genic sequences is low (Bryan et al., 1999), but may be useful genetic variation for exploitation in breeding programmes.
f) SNPs are menable to non-gel based assays, hence very high throughput can be achieved.

g) STMS or STR scoring is platform dependent (Baric et al., 2008) and more prone to genotyping errors (Hoffman et al., 2005)

Furthermore, a recent study on the comparison of SNPs and microsatellites as linkage mapping markers by Ball et al., 2010 have shown that when using multiple types and large numbers of markers to create dense linkage maps, the least error prone loci (i.e. SNPs) rather than the most informative ones should be used to create framework maps before the addition of other potentially more error prone markers (microsatellites). Glover et al., 2010 also found that best 15 SNPs (30 alleles) gave a similar level of self-assignment compared to the best 4 STR loci (83 alleles); however, addition of further STR loci did not lead to a notable increase assignment whereas addition of up to 100 SNP loci increased assignment of individuals to populations. However, because allele frequencies can vary greatly among populations, it is important to validate and identify the informative SNPs that are useful for sample identification in biomedical and epidemiological studies. We have now screened a total of 28 SNPs based mostly on the SNPs chosen by an earlier work conducted by Lee et al., 2005, which was carried out for individual identification in a Korean population in which one SNP has been selected for each of the chromosomes including sex chromosomes using the single-base extension method with SNaPshot™ Kits (Applied Biosystems). Since the project conducted by Lee et al., (2005) focused on three major populations (Caucasians, Asians and African-Americans), and they chose SNPs that were thought to be highly informative for all three populations, we hypothesized that they would also be appropriate for our study on Chinese subjects from Hong Kong. In this study, we also attempt to lower the cost of genotyping by using the iplex method of Sequenom. However, we found in the present study, some of the SNPs (2/22) used in Lee's paper were not suitable for the Hong Kong Chinese samples (384 pairs of twins) we analyzed. In order to compensate for the reduction in the number of useful SNPs, we have added in four SNPs (of which 3 are independent) and these were found to be useful for our samples. In this report, we present the results obtained using a combination of the SNPs published in Lee et al., 2005, plus the 6 SNPs we chose, on their effectiveness in distinguishing identical and non-identical twins (768 individuals) for the Hong Kong Chinese population using the iplex method of Sequenom. We analyzed the inter-dependence of these markers and we also validated the Sequenom method using the microsatellite method using a subset of the samples.

MATERIALS AND METHODS

SNP Typing and DNA Extraction

The markers were chosen according to the following criteria:

Other than the SNPs from the paper conducted Lee et al., 2005 on individual identification in Korean population, 4 other SNPs were chosen based on our previous experience that they work well in the Sequenom platform. We did not use 2 SNPs: rs 1820795, rs 746018 from Lee et al., 2005, to make room for some SNPs that we had genotyping data for some positive control DNAs. As much as possible, the following rules for selection of SNPs in the Twin Study were followed: For each allelic SNP,

1. Allele frequency close to 50:50 allelic distribution are preferred
2. Minor genotype >5% in Han population
3. Avoid selection of SNPs at genome position with copy number variation
4. Published SNPs of HapMap project obtained by means of Sequenom were preferred.

DNA was extracted according to the instruction manual from OraGene (DNA Genotek Inc, Ontario, Canada) and the amount sent out for genotyping by Sequenom is 600ng with concentration of 20ng/uL. Although we sent out 600ng of DNA for our experiments, the actual amount of starting material required for Sequenom analyses can be as low as 135ng (at a minimum concentration of 5ng/ul), according to specification of iplex DNA required for customer service at the Broad Institute (http://www.broadinstitute.org/science/projects /broad/ncrr-center-genotyping-analysis/dna-sample-quality-and-quantity-specification). The average yield of the DNA was 6.85 mg from 2 ml of saliva with an average concentration of 48ng/ul. DNA samples that had lower than 20ng/ul were concentrated further. The A260/A280 ratio of 1.7 – 2.0 was obtained using UV spectrophotometer. A representative sample was used in PCR reaction to confirm that there were no inhibitors in the DNA samples.

DNA Genotyping by the MALDI-TOF Mass Spectroscopy Method Using Sequenom

SNP genotyping was performed *via* customer service using the multiplex technology of Sequenom offered by McGill University and Genome Quebec Innovation Centre. The Sequenom iPlex Gold technology is ideal for medium sized projects where scoring between 5 and 300 SNPs on hundreds to a few thousands of DNA samples is needed.

In the first experiment, a total of 674 DNA samples with 8 control DNA samples (having previously typed genotypes of 4 SNPs) were aliquoted onto 8 microtiter plates and the purified DNAs were delivered by Express service at room temperature. 28 plex oligonucleotide primers were designed and iPLEX Gold assay on the MassARRAY® Platform were performed. A second set of 94 samples were genotyped by Sequenom, including 44 samples that failed the genotyping by microsatellite method as well as to 18 randomly drawn heterzygotic and 14 monozygotic samples from those successfully typed by microsatellite test.

SNPs Analyzed

Genotyping for twenty-eight SNPs were performed and their allele frequencies determined. The Chromosomal locations were updated by using data from the HapMap Data Release 27 (containing Phase II + Phase III data) updated on Feb 2009, on NCBI B36 Assembly. Most SNPs location were similar to earlier version (as reported by Lee et al., 2005) with the exception of some SNPs that has a more significant change, e.g. rs264729 was at 13:16,177,910 while the newer location was assigned to be at 13:21,095,910 and rs879835 was at 19:31,750,866 while the new location was assigned to be at 19:35,993,726, rs922645 was at 14: 50,268,752 while the new location was assigned to 14:55,542,468.

Analytic Methods

Allele frequencies for each marker were estimated by counting the alleles within our population assuming each marker to be a two-allele system (except for the sex chromosome markers). Agreement with Hardy-Weinberg ratios was tested for each marker using a simple Chi-square test comparing the expected and observed number of individuals occurring for each possible genotype. The LD values were then examined in various ways for any associations among the markers. The match probability that two unrelated individuals will have the same multi-locus genotype was calculated as described previously (Kidd et al., 2006). The frequency of the best 26 loci genotype was calculated assuming Hardy-Weinberg ratios and the independence of the loci.

Confirmation of Identity Status by Microsatellite Method

The ABI microsatellite kit (AmpFISTR identifiler PCR amplifcation kit) for measuring 16 loci was used to confirm the status of those marginal cases (8 pair of twins Cases) with more than 1 missing data from the best 24 SNPs panel. It was found that in two cases where there were missing data (1 or 2 missing SNPs) for the SNP test, there were also missing peaks from the AmpFISTR identiflier kit though both are probably identical twins as indicated by the large number of shared alleles. Thus the missing SNPs might be due to structural changes of genomic DNA or sample inhibition of PCR. Overall, we found that if missing data were discounted in the 8 pairs of twins that had missing data, the prediction using Sequenom and AmpFISTR methods concur in 100% of the cases. Further test with 18 pairs of dizygotic twins and 24 pairs of identical twins that were randomly drawn from the microsatellite experiments and then re-tested with the Sequenom method were shown to be all concordant with the microsatellite method. Though we found that there were more failed reactions using the microsatellite method compared with the SNP method, we cannot rule out the possibility that if we perform more PCR cycle, the sensitivity of the microsatellite method could be increased.

RESULTS

Allele Frequencies

Of the 28 SNPs that were genotyped (listed in Table 1), three SNPs (marked with * in Table 1) had missing data at higher than 2% frequency. Fisher's exact tests were performed to evaluate compliance with the Hardy–Weinberg equilibrium. Fisher's exact test showed that all markers are in Hardy–Weinberg equilibrium ($P > 0.05$). Of the 25 best SNPs that were genotyped, 16,834 (674*25-16) typings succeeded and 0.095% (16/16,834) was missing/failed assays in the first experiments. 99.85% (673/674) of the individuals completed 24 of the 25 SNPs in the first experiments.

Table 1. Summary results of the SNPs used for genotyping

SNP no.	SNP ID.	Chromosome location	No. missing/failed SNPs (%). The split cell shows failed results from 2 experiments.	Source@ (allele frequency-calculated from 674 samples only)	Chromo-some
1	rs1009480^	34,512,183	-- 1/94 (1%)	1 (43:57) 2 (49:51) 3 (47:52)	3
2	rs1014803^	36,015,910	--	1 (47:53) 2 (43:57) 3 (50:50)	21
3	rs1158810^	120,849,340	--	1 (43:57) 2 (47:53) 3 (50:50)	9
4	rs13397241	75,715,708	1/674(0.15%) --	2 (46:54) 3 (45:55) 4 (45:55)	2
5	rs1346081^*	67,717,314	43/674 (6.4%) 9/94 (9.6%)	1 (48:52) 2 (48:52) 3 (43:57)	4
6	rs138000^	38,684,927	--	1 (40:60) 2 (40:60) 3 (37:63)	22
7	rs1474454^	36,045,007	--	1 (45:55) 2 (44:56) 3 (39:61)	17
8	rs1821380^	37,100,694	--	1 (33:67) 2 (31:69) 3 (39:61)	15
9	rs205119*^	25,576,403	115/674 (17%) ND	1 (42:58) 2 (48:52) 3 (35:65)	16
10	rs2143340	24,767,050	--	2 (13:87) 3 (15:85)	6
11	rs215695^	32,364,433	--	1 (50:50) 2 (46:54) 3 (50:50)	7
12	rs2196155^	21,074,650	--	1 (13:87) 2 (23:77) 3 (6:94)	Y
13	rs264729^	21,095,910	--	1 (58:42) 2 (38:62) 3 (40:60)	13
14	rs298612^	130,985,367	--	1 (45:55) 2 (50:50) 3 (49:51)	8
15	rs3756821	24,754,800	--	2 (41:59) 3 (24:76) 4 (20:80)	6

SNP no.	SNP ID.	Chromosome location	No. missing/failed SNPs (%). The split cell shows failed results from 2 experiments.	Source@ (allele frequency-calculated from 674 samples only)	Chromo-some
16	rs4774768*	53,513,797	16/674(2.4) 4/94 (4.3%)	2 (24:76) 3 (24:76) 4 (16:84)	15
17	rs525218^	22,665,111	--	1 (57:43) 2 (50:50) 3 (48:51)	18
18	rs719569^	38,194,421	4/674(0.59) 0/94 (<1%)	1 (43:57) 2 (39:61) 3 (42:58)	10
19	rs730123^	11,992,879	--	1 (47:53) 2 (54:46) 3 (48:51)	1
20	rs752761^	158,864,926	--	1 (45:55) 2 (47:53) 3 (41:59)	6
21	rs7543044	35,446,898	--	2 (48:52) 3 (47:53) 4 (45:55)	1
22	rs807507	24,687,846	1/674(0.15%) 1/95 (1%)	2 (18:82) 3 (20:80) 4 (20:80)	6
23	rs879835^	35,993,726	--	1 (58:42) 2 (48:52) 3 (40:60)	19
24	rs880537^	93,011,146	2/674(0.3 %) 0/94 ((<1%)	1 (37:63) 2 (50:50) 3 (44:56)	12
25	rs911135^	18,609,293	--	1 (47:53) 2 (39:61) 3 (36:64)	20
26	rs921115^	45,095,120	--	1 (38:62) 2 (46:54) 3 (42:58)	2
27	rs922654^	55,542,468	1/674(0.15%) 0/94 (<1%)	1 (45:55) 2 (43:57) 3 (36:64)	14
28	rs997262^	95,132,016	6/674(0.89%) 3/94 (3%)	1 (50:50) 1 (53:47) 2 (53:47) 3 (45:55))	X

-- absence of failed reaction.

^ SNPs chosen accordingly to Lee et al., 2005.

*overall missing data rate less than 2%.

@ Source of allele frequencies- 1: Lee et al., 2005; 2: Gibb et al., 2003; Frazer et al., 2007 (3) or dbSNP web page; 3: This study (Data from the first dataset of 674 samples only which might be somewhat biased due to relatedness of the 674 samples); 4: Unpublished data from our previous study with a different sample set that are not twins.

Table 2. Inter-dependence of the SNP markers

Number	SNP	Chromosome location	Linked SNPs
1	rs1009480	3: 34,512,183	
2	rs1014803	21: 36,015,910	
3	rs1158810	9: 120,849,340	
4	rs13397241	2: 75,715,708	
5	rs138000	22:38,684,927	
6	rs1474454	17: 36,045,007	
7	rs1821380	15:37,100,694	
8	rs2143340*	6: 24,767,050	+++
9	rs215695	7: 32,364,433	
10	rs2196155	Y: 21,074,650	
11	rs264729	13: 21,095,910	
12	rs298612	8: 130,985,367	
13	rs3756821	6: 24,754,800	+++
14	Rs4774768	15: 53,513,797	
15	rs525218	18:22,665,111	
16	rs719569	10: 38,194,421	
17	rs730123	1: 11,992,879	
18	rs752761	6:158,864,926	
19	rs7543044	1: 35,446,898	
20	rs807507	6: 24,687,846	
21	rs879835	19: 35,993,726	
22	rs880537	12: 93,011,146	
23	rs911135	20: 18,609,293	
24	rs921115	2: 45,095,120	
25	rs922654	14: 55,542,468	
26	rs997262	X:95,132,016	

Under SNP pair column, the numbers in front of each pair correspond to the row number in table 2.
*Either no. 8 or no. 13 should be used.

In the second experiments with a lower number of samples (94 samples), the results obtained were similar (see bottom row of split cells in Table 1).

Inter-Dependence of the Best 26 SNPs

As shown in Figure 1, the best 26 SNPs (25 best SNPs plus rs4774768) were analyzed to assess the population independence of variation for the SNPs. Genotyping data from one individual of each pair of twins was selected. Linkage disequilibrium (LD) block and Hardy-Weinberg test were also done by Halpoview (Barrett et al., 2005). The vast majority of the markers were independent (except for markers rs807507 and rs3756821 which are both markers for the same gene (KIAA0319) and thus acts as a positive controls for this test)-Table 2. As shown by our analyses of inter-dependence of SNPs, either no. 8 or no. 13 should be used in the panel for our population due to linkage (see Figure 1 and Table 2), and since no. 13 has a higher minor allele frequency, it is recommended that no. 8 can be dropped in the final panel used for Hong Kong Chinese population.

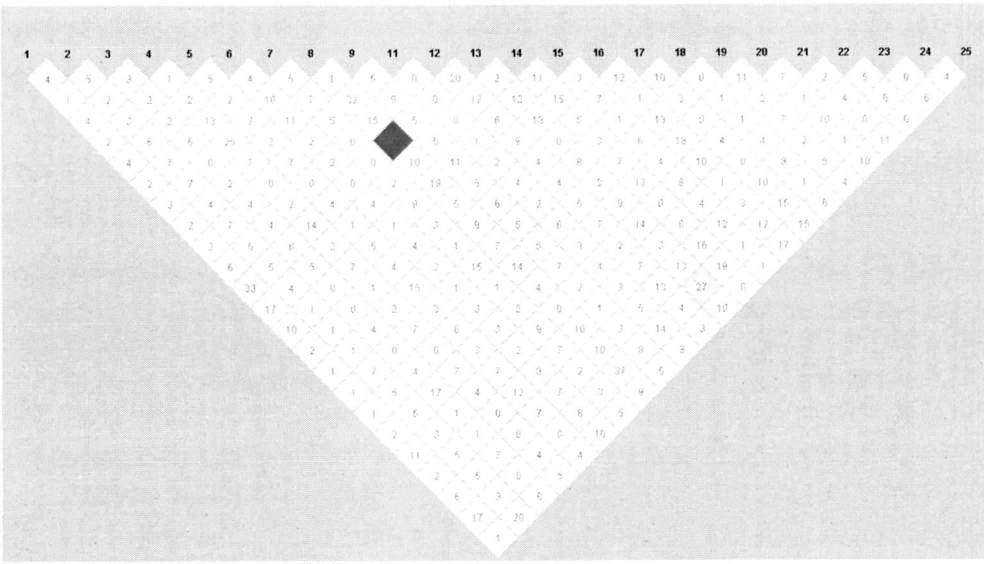

Figure 1. Linkage disequilibirum analyses by using Haploview: LD are displayed with recombination rate plotted above (Barrett et al., 2005).

Statistics of the 28 SNP Panel

The frequencies of the dizygotic differences in genotypes of the 28 SNP panels are represented graphically (Figure 2). For the dizygotic twins (i.e. a total of 180 individuals or 90 pairs of twins), the frequency distribution of the cumulative number of different SNPs (setting a cut-off value of 3 due to the 3SNPs that had high failure rate) is represented in Figure 2. The average is approximately 7.2 different SNPs while the maximum number of different SNPs is 13 (Figure 2).

For population data, the gene count method was used to calculate observed allele frequencies and the observed genotypes at each of the 24 SNP loci. P_I was defined as the estimated probability that two unrelated individuals selected at random would possess identical multi-locus genotypes and was calculated for each SNP from genotype frequencies as previously described (Jones, 1972 and Holt et al., 2000). For parentage testing, P_E, defined as the probability that a random candidate male would be excluded from paternity with a known maternal genotype, was calculated for each SNP from allele frequencies, as previously described (Chakraborty R. and D.N. Stivers, 1996). The calculation of P_I at X chromosomal SNP was carried out as described by Desmarais et al., 1998.

The Use of Excel Macros to Determine the Zygosity Status

An Excel worksheet with macro function to automatically call whether the twin is dizygotic or monozygotic is constructed as users fill in the sample information.

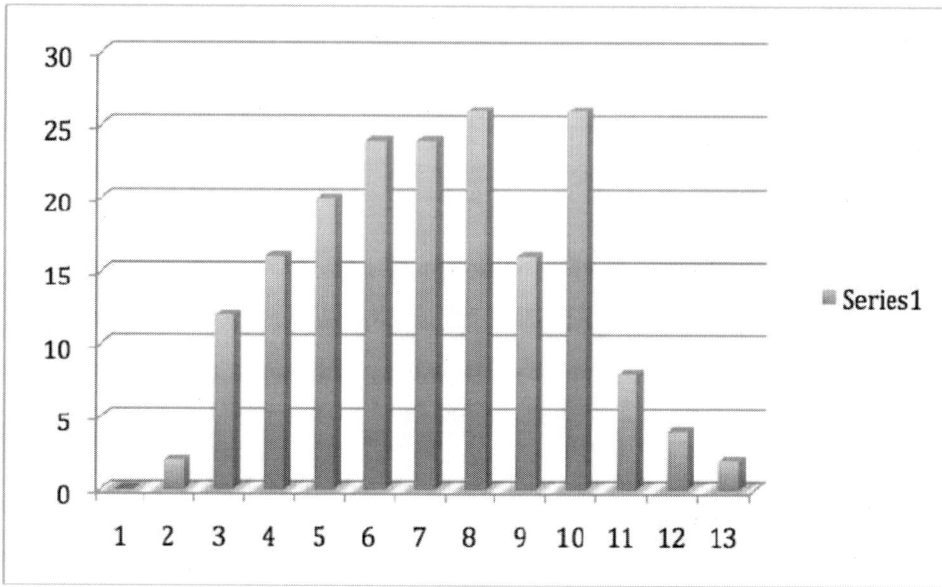

Figure 2. The differences in SNP genotypes for the dizygotic twins. The cumulated occurrence of different SNPs for the dizygotic twins is plotted along the Y axis while the observed number of different SNPs for the dizygotic twins are plotted in the X axis.

Input is allowed with different number of samples, as well as different cut off number of minimal number of SNPs before dizygosity is called. Choice of chi square value (depending on the p-value one wish to set) is also allowed. An option to mask SNPs that fail Hardy-Weinberg test is also allowed. The output of total number of DZ and MZ and total number of SNPs that failed HW test is displayed on the "Input" sheet.

Discussion

In terms of the SNPs chosen for the study of Chinese from Hong Kong, this study represents the first study to date to find SNPs for identification purpose for this population. The final panel of 25 SNPs is appropriate for use in human identification. We conclude that the 25 SNPs in our final panel are statistically independent in our population. The median and mean LD values are close to zero and the computed LD values are significantly different from zero and close to what would be expected by chance only. About 99% of all LD values are less than or equal to 0.25 and they were mostly different SNP pairs located on different chromosomes (19 SNPs were pairs from different chromosomes) or are far apart on the same chromosome (Only one SNP pair was less than 50 MB apart on the same chromosome which serves as a positive control for the test on Linkage Disequilibrium). The reasons that two SNPs from Lee et al., 2005, had higher missing data rate in our study could be due to variation between Koreans and Chinese (in Hong Kong) in the sequences where the primers bind, and if such variation lead to mismatches between the primer binding sites and the template DNA strand, then those individuals will give missing signals in the sequenom genotype analyses. These missing data could also be due to the difference in platform used:

the technology used by Lee et al., 2005 is the single-base extension method with SNaPshot™ Kits (Applied Biosystems) whereas the platform used in this study is the multiplex (iPLEX) method of Sequenom. In total, we have performed genotyping for 76 samples (or 38 pairs of twins) with both microsatellite method and the Sequenome method and the results were all concordant.

Discrimination among Individuals

Our panel of 25 SNPs resulted in grouping the samples into identical twins or non-identical twins. We obtained the nearly symmetric distribution of around 50% of matches for the non-identical twins (centered around 7 mismatches out of 25 genotypes-figure 2). According to our results, 1-2 missing data points in the 24 SNP profiles has a high probability of being an error and should be replicated whereas any mismatch (even one single one) other than missing/failure data would suggest that the individuals are non-identical twins. This suggests that the marker set is appropriate for distinguishing identical and non-identical twins in genetic studies.

This same set could also be used in paternity assays, and according to our study, the single worse performing pair of twins had information from 22 autosomal SNP and according to the paper by Lee et al., 2005, the estimated probability of identity (P-I) and probability of paternity exclusion (P-E) with 22 autosomal SNP loci were 1.905 x 10(-10) and 98.9%, respectively. Such exclusion probability is not impressive compared with that obtained by the microsatellite method (and STRs form the basis of dozens of National DNA Databases which are comprehensive in scope and an essential crime-lab tool); however, the saving of approximately 10-20 fold in the cost of SNP analyses compared to the microsatellite method indicates that SNP analyses using iPLEX method of Sequenom is clearly preferable in this context (Table 3).

Table 3. Comparison of two (STR and SNP) technologies for zygosity testing

Features	Time needed for analyses	Power in exclusion	Minimal Cost per sample (HKD)#	Cost for reaction	Other considerations
Microsatellite method (STR)	Longer*	Higher**	~$1,100*	$22,000/200 samples	Fading of fluorescent dye is a factor in deciding when/where samples should be sent for electrophoresis.*
SNP (iPLEX by Sequenom)	Shorter**	Lower*	~$50**	Approximately $35,000/700 samples	Automation more easy**

**More desirable features in this context.

*Less desirable feature in this context.

(excluding shipping costs for customer service, equipment cost, software needed to interpret results, and cost of preparation of DNA samples, repeating failed samples, and estimation of DNA concentrations).

CONCLUSION

Our results suggest that there are similar SNPs that are useful for identification purpose between Koreans and Southern Chinese, however, some SNPs (approximately 9% or 2/22 SNPs from Lee et al's paper that were tested) have noticeable differences. Our observation is consistent with the conclusion of Jin et al., 2009, who recently reported that the peopling of Korea can be seen as a complex process, interpreted as an early northern Asian settlement and they suggested at least one subsequent male-biased southern-to-northern migration, possibly associated with the spread of rice agriculture. Thus some differences between the Southern Chinese and Korean were not unexpected. However we cannot compare the results of the our studies with Jin et al's data directly because they analyzed mitochondrial DNA and Y chromosome STR data to elucidate the genetic structure of East Asian populations whereas we used SNP data from mostly autosomal region of Southern Chinese.

Furthermore, since our samples are all twins, we cannot rule out the possibility that the observation of the allele frequencies of our sample is biased due to pre-selection of samples that favour the likelihood of bearing twins or seeking in vitro fertilization therapy who are more likely to bear twins. In conclusion, we have developed a panel of 25 SNPs and a strategy for analysis that are useful for distinguishing monozygotic twins from dizygotic twins in a high-throughput and relatively simple manner compared to the usual microsatellite approach for forensic studies.

ACKOWLEDGMENT

The authors wish to acknowledge financial support from the Research Grants Council(HKU743907H), and the Croucher Foundation. TL Yeung is a recipient of UGC studentship. The help of Dorothy Bioshop of Department of Psychiatry, Oxford University, UK, in reading the manuscript is much appreciated. The help of Aaron Sung in the macro worksheet is much appreciated.

REFERENCES

Angers, B; Estoup, A; Jarne, P. Microsatellite size homoplasy, SSCP, and population structure: a case study in the freshwater snail Bulinus truncatus. *Mol. Biol. Evol.* 2000; 17:1926-1932.

Ball AD, SJ; Dawson, DA; Birkhead, TR; Burke, T; Slate, J. A comparison of SNPs and microsatellites as linkage mapping markers: lessons from the zebra finch (Taeniopygia guttata). *BMC Genomics.* 2010; 2010: 218.

Baric S, MS; Hofer, M; Grill, D; and Dalla, Via J. Comparability of genotyping data obtained by different procedures an inter-laboratory survey. *Journal of Horticultural Science and Biotechnology.* 2008; 83:183-190.

Barrett, JC FB; Maller, J; Daly, MJ. Haploview: analysis and visualization of LD and haplotype maps. *Bioinformatics.* 2005; 21:263-265.

Bhatia, VK; Gupta, VK; Bhar, Lal Mohan. *Advances in Data Analytical Techniques.* 2001.

Bryan GJ, SP; Collins, A; Kirby, J; Smith, JB; Gale, MD. Low levels of DNA sequence variation among adapted genotypes of hexaploid wheat. *TAG Theoretical and Applied Genetics.* 1999; 99:192-198.

Chakraborty, R; Stivers, DN. Paternity exclusion by DNA markers: effects of paternal mutations. *J Forensic Sci.* 1996; 41:671-677.

Colson, I GD. Evidence for complex mutations at microsatellite loci in Drosophila. *Genetics.* 1999; 152:617-627.

Consortium, IH. The International HapMap Project. *Nature.* 2003; 426:789-796.

Desmarais, D ZY; Chakraborty, R; Perreault, C; Busque, L. Development of a highly polymorphic STR marker for identity testing purposes at the human androgen receptor gene (HUMARA). *J. Forensic Sci.* 1998; 43:1046-1049.

Frazer, KA; et al. A second generation human haplotype map of over 3.1 million SNPs. *Nature.* 2007; 449:851-861.

Glover KA, et al. A comparison of SNP and STR loci for delineating population structure and performing individual genetic assignment. *BMC Genet.* 2010; 11:2.

Hoffman, JI; Amos, W. Microsatellite genotyping errors: detection approaches, common sources and consequences for paternal exclusion. *Mol. Ecol.* 2005; 14:599-612.

Holt, CL; et al. Practical applications of genotypic surveys for forensic STR testing. *Forensic Sci Int.* 2000; 112:91-109.

Jin, HJ; Tyler-Smith, C; Kim, W. The peopling of Korea revealed by analyses of mitochondrial DNA and Y-chromosomal markers. *PLoS One.* 2009; 4:e4210.

Jones; DA. Blood samples: probability of discrimination. *J. Forensic Sci Soc.* 1972; 12:355-359.

Kidd, KK; et al. Developing a SNP panel for forensic identification of individuals. *Forensic Sci. Int.* 2006; 164:20-32.

Kwok, PY; Deng, Q; Zakeri, H; Taylor, SL; Nickerson, DA. Increasing the information content of STS-based genome maps: identifying polymorphisms in mapped STSs. *Genomics.* 1996; 31:123-126.

Lee, HY; et al. Selection of twenty-four highly informative SNP markers for human identification and paternity analysis in Koreans. *Forensic Sci. Int.* 2005; 148:107-112.

In: Biochemistry and Molecular Biology Coming of Age
Editor: Mary Miu Yee Waye
ISBN: 978-1-63321-815-4

Chapter 4

THE DYSLEXIA CANDIDATE GENE KIAA0319L ENCODES N-GLYCOSYLATED ISOFORMS THAT FORM HOMO-DIMERS

Ming-Wai Poon, Hoi-Ling Chan, King-Poo Lim and Mary Miu-Yee Waye*
School of Biochemical Sciences, Laboratory for Human Genomics, The Chinese University of Hong Kong, Shatin, N.T., Hong Kong, China

ABSTRACT

KIAA0319-Like (KIAA0319L) gene has been reported to be associated with developmental dyslexia. It has been suggested to be a functional and positional candidate for developmental dyslexia, as it is a homologue of *KIAA0319* and positioned near the developmental dyslexia susceptibility locus, DYX8. This article documents the expression pattern of this gene in various human tissues, including the adult and fetal brain, by quantitative reverse-transcriptase polymerase chain reaction (RT-PCR). *KIAA0319L* mRNA was present in both adult and fetal brain, and was found to be most abundant in the cerebellum. Kiaa0319L protein was shown to be mainly localized in the cytoplasm, the periphery of the nucleus and also within the nucleus. Experiments with overexpressed constructs containing the Kiaa0319L protein revealed that the full-length Kiaa0319L protein is an N-glycosylated protein which exists in at least 2 glyco-isoforms. This protein was able to bind to itself in a co-immunoprecipitation experiment. These properties of Kiaa0319L protein are also similar to its homologue, the Kiaa0319 protein, which is the N- and O- glycosylated plasma membrane isoform A studied by an Italian group. We report here a novel study comparing the two Kiaa proteins, their protein properties and predicted domain structures. This is an important initial step to the functional studies of the two Kiaa proteins.

Keywords: Kiaa0319L, Kiaa0319, dyslexia, protein, brain

---

* Corresponding author: Prof. Mary Waye, School of Biomedical Sciences, Croucher Laboratory for Human Genomics, The Chinese University of Hong Kong, Shatin, N.T., Hong Kong, China. Tel:852-26096874 Fax:852-26037732; E-mail: mary-waye@cuhk.edu.hk.

Introduction

Developmental dyslexia (DD) a common cognitive disorder affecting around 5-17% of school-aged children [1], often arises from impaired phonological awareness. It is defined as a persistent difficulty in learning to read that is not explained by sensory deficits, lack of motivation, or lack of adequate reading instruction.

DD is strongly heritable, occurring in up to 50% of individuals who have a parent or sibling with DD and 68% of identical twin. Genetic studies have identified nine regions on the genome linked to DD; these regions are named DYX1-9. These genetic factors are distributed on chromosome 1,2,3,6,11,15,18 and x with their associated candidate susceptibility genes *DCDC2* [2, 3], *DYX1C1* [4, 5], *KIAA0319* [6-9], *ROBO1* [10], *MRPL19, C2ORF3* [11] and *KIAA0319L* [12]. Among them, *KIAA0319L* showed a modest significance for a single marker (rs7523017; P=0.042). It has been suggested to be an intriguing positional and functional candidate for DD, this gene is located within linkage peaks from three independent studies on chromosome 1p (Rabin et al., 1993; Grigorenko et al., 2001; Tzenova et al., 2004), making it a suitable positional candidate whereas the homology to Kiaa0319 protein encoded by *KIAA0319,* a strong DD candidate gene, on 6p suggests that it is a potential functional candidate. (Couto et al., 2008) [6, 9, 13]. We have undertaken a functional analysis of Kiaa0319L protein to understand the molecular mechanisms this protein may be involved in with regard to brain functioning. In this context, we have shown a specific distribution of the Kiaa0319L protein in the cortex, cerebellum and hippocampus which are regions known to be involved in reading tasks and DD [6, 14-17]. Our previous data screened Kiaa0319L's interacting partners. In yeast-two-hybrid and co-immunoprecipitation assay, our previous data provides evidence for Kiaa0319L interacting with Nogo Receptor 1. This supports the idea that Kiaa0319L participates in axon guidance [18]. The functional characterization of Kiaa0319L and that of Kiaa0319 proteins, in turn, would allow us to gain new insights into their biological functions and the biological mechanisms underlying the reading process such as neuronal migration and axon guidance. Here, we present an analysis of the protein properties of Kiaa0319L.

Materials and Methods

Synthesis of cDNA from total RNA of 20 human tissues and primers for RT-PCR. Total RNA from 20 human tissues in Human Total RNA Master Panel II (Clontech) were used to synthesis first-strand cDNAs from 1μg of total RNA in a total volume of 20μl using the SuperScript™ III First-Strand Synthesis SuperMix (Invitrogen,USA) according to the manufacturer's protocol.

Plasmids. cDNA of Kiaa0319L isoform a (NM 024874) were cloned into pCMV-myc and pEGFPC1 to generate pCMV-myc-Kiaa0319L and pEGFPC1-Kiaa0319L for constitutive expression (Table 1).

Antibodies. Custom rabbit polyclonal Kiaa0319L antiserum specificity has been previously described [18, 19]. The following additional antibodies were used, (Primary antibodies) mouse monoclonal anti-myc tag (9B11) antibody (Cell Signaling Technology, Beverly MA); mouse monoclonal IgG2a anti-GFP tag (B-2) antibody (Santa Cruz

Biotechnology. Inc., California) and mouse monoclonal IgG1 anti-beta actin (AC-15) antibody (Sigma-Aldrich, Inc.), (Secondary antibodies) goat anti-rabbit IgG HRP and rabbit anti-mouse IgG HRP (Dako, Denmark), (Secondary antibodies for histochemical staining) donkey anti-rabbit fluorescein isothiocyanate (FITC) (Jackson ImmunoResearch, USA).

Table 1. Primers used in cloning human full-length Kiaa0319Lprotein into pCMV-myc and pEGFP-C1 expression constructs

cDNA	Primer sequences
Kiaa0319L pCMVmyc	F: 5'-TAGGGCGTCGACCATGGAGAAGAGGCTGGGAGT-3'
	R:5'-TAGGGCGGTACCTTACTACAGGATCTCCTCCCGCG-3'
Kiaa0319L pEGFPC1	F:5'-TAGGGCGTCGACATGGAGAAGAGGCTGGGAGT-3'
	R:5'-TAGGGCGGTACCTTACTACAGGATCTCCTCCCGCG-3'

Note: the full length of Kiaa0319L isoform a cDNA encodes 1049 amino acids.

Table 2. RT-PCR Primer pairs used for the expression analysis of the KIAA0319L transcript, KIAA0319L Variant 1

Gene	Primer Sequence (5'to3')	Size of Amplicon
Hydroxymethylbilane synthase (*PBGD*)	Forward Primer: CCATGTCTGGTAACGGCAAT Reverse Primer: GTCTGTATGCGAGCAAGCTG	106 b.p.
KIAA0319L	Forward Primer:GAGCACCTCAGCACCATACC Reverse Primer: GGCAGGGTTATCTGGACACT	75 b.p.

SYBR® Green real time PCR. Total RNA was reverse transcribed into first strand cDNA using SuperScript® III First-Strand Synthesis Kit (Invitrogen). SYBR® Green real time PCR was conducted in triplicate. The amplifications were performed as follows, 2 mins at 50°C, 10 mins at 95°C and then 40 cycles each at 95°C for 15 sec and 60°C for 60 sec in the 7500 Fast-Real-Time PCR System. For KIAA0319L, the forward and reverse primers which are listed in table 2 were used. Results were normalized to an internal control transcript that encodes hydroxymethylbilane synthase (PBGD); the forward and reverse primers are listed in table 2.

Cell culture and transfection. Human embryonal kidney cells (HEK-293FT) (Invitrogen, USA) and human neuroblastoma cells (SH-SY5Y) (ATCC, USA) were cultured in Dulbecco's modified Eagle's medium/ F12 1:1 (Hyclone) (DMEM) supplemented with 10% (v/v) fetal bovine serum and 100mg/ml penicillin/streptomycin. HEK cells were transfected using Lipofectamine™ Transfection Reagent (Inivitrogen) and PLUS™ Reagent (Invitrogen). A total of $1.5x10^5$ cells were seeded in T-75 flasks (in triplicate) and propagated. When cells reached ~80% confluence, they were seeded in 100mm dishes for transfection.

Protein interaction studies and immunohistochemical staining. For immunoprecipitation, HEK cells were lysed after about 48hours post-translation in RIPA buffer with protease inhibitors. Cells were lysed for 30 mins at 4°C, and cleared by centrifugation. Supernatants were pre-cleared and incubated for 1hour with antibody cross-linked to protein G-beads (GE Healthcare). After washing with lysis buffer, bound proteins were eluted in Laemmli buffer and analyzed by immunoblotting. For immunoblotting, cells were lysed in RIPA buffer

(50mM Tris-HCl, pH=8, 150mM NaCl, 1% Triton X-100) with protease inhibitors (Roche). Immunoblots were developed by chemiluminescence (Perkinelmer).

Histochemical staining. SH-SY5Y cells were plated onto glass coverslips. 18hrs after the cells were seeded, cells were fixed with 3.7% formaldehyde, permeabilized with Triton X-100, and stained with primary antibody anti-Kiaa0319L antibody (1:100) followed by FITC-labeled donkey anti-rabbit secondary antibody (Jackson ImmunoResearch, USA). The cells were then stained with propidium iodide, (PI) (10μg/ml) with RNase A(100μg/ml) in phosphate buffered saline solution (PBS) to visualize the nuclei.

RESULTS

Expression of *KIAA0319L* Transcript in Human Tissues and Kiaa0319L Protein in Human Neuroblastoma Cell Line SH-SY5Y

KIAA0319L mRNA has a broad tissue distribution. *KIAA0319L* mRNA was detected in the heart, the kidney, the lung, the placenta, the prostate and the thyroid gland. However, it is more abundantly expressed in adult and fetal brain, especially in the cerebellum, which has the highest expression. Our results suggest that *KIAA0319L* has very important roles in the brain, but there might also be other house-keeping functions as it is nevertheless present in other body tissues (Figure 1A).

A

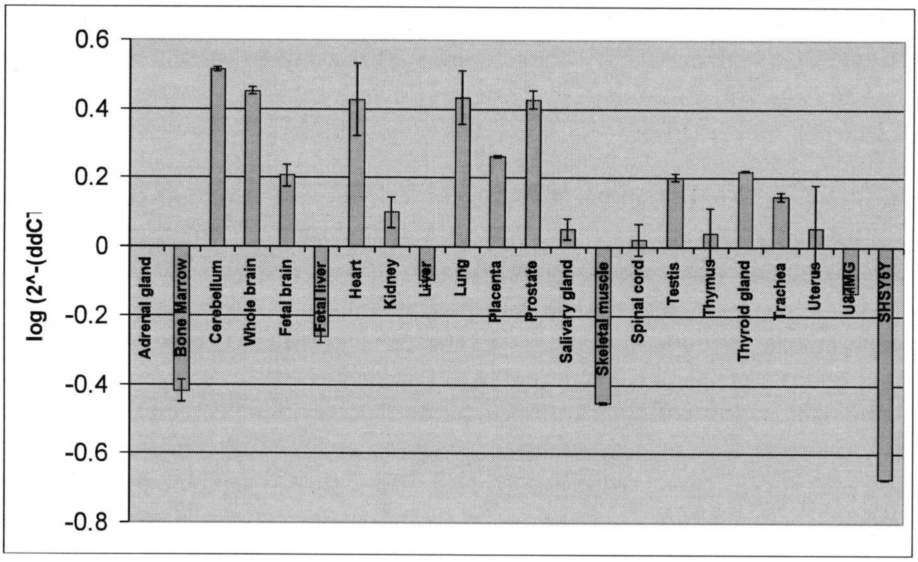

Figure 1A. *KIAA0319L* expression in human tissues and cell lines U87MG and SH-SY5Y. Tissue distribution of *KIAA0319L* in human tissues. *KIAA0319L* mRNA level in the adrenal gland is defined as zero.

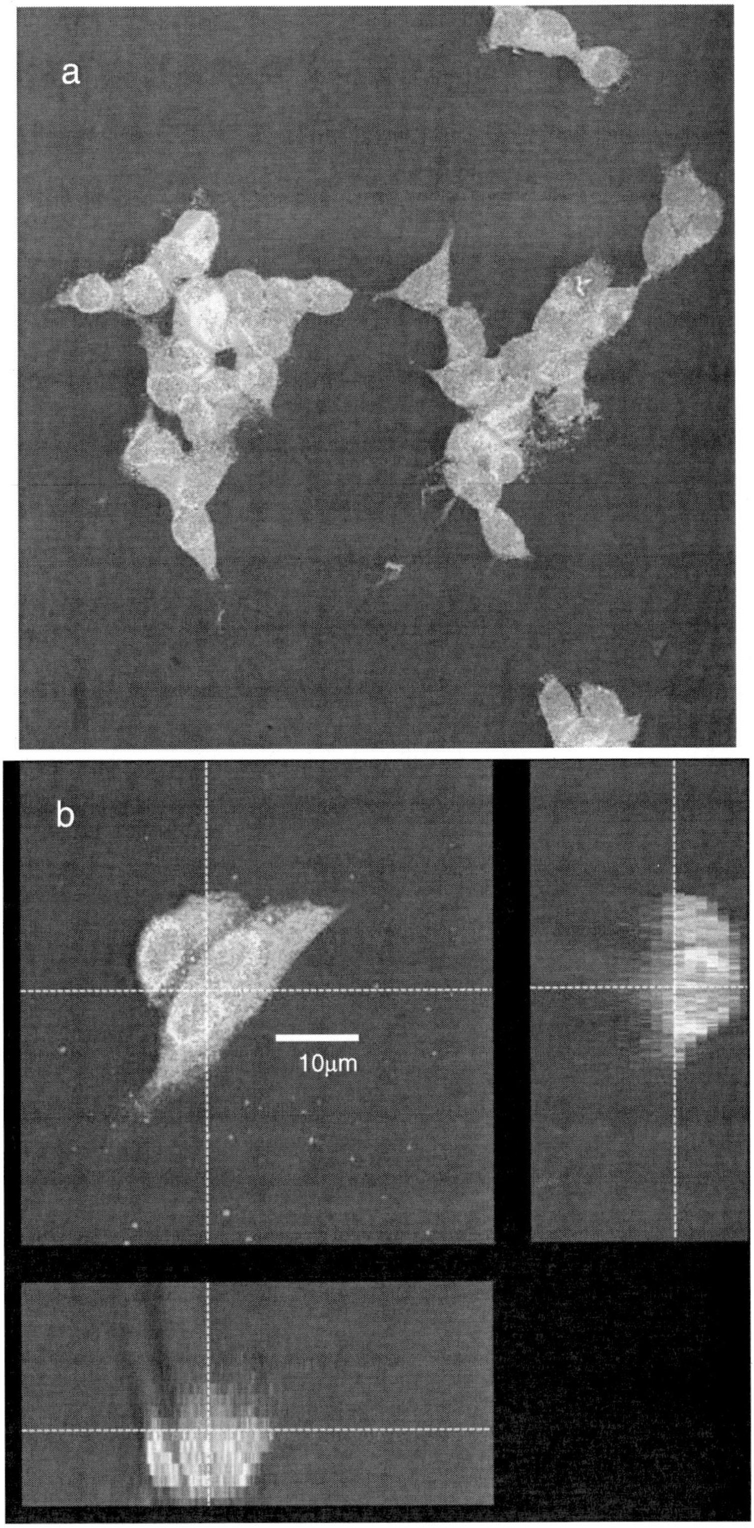

Figure 1B. (Continued).

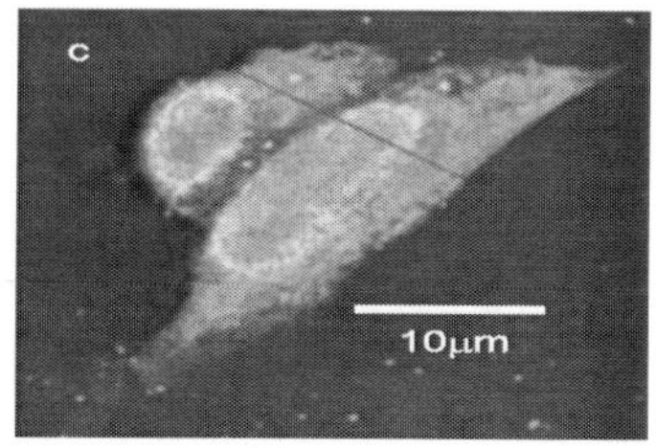

d

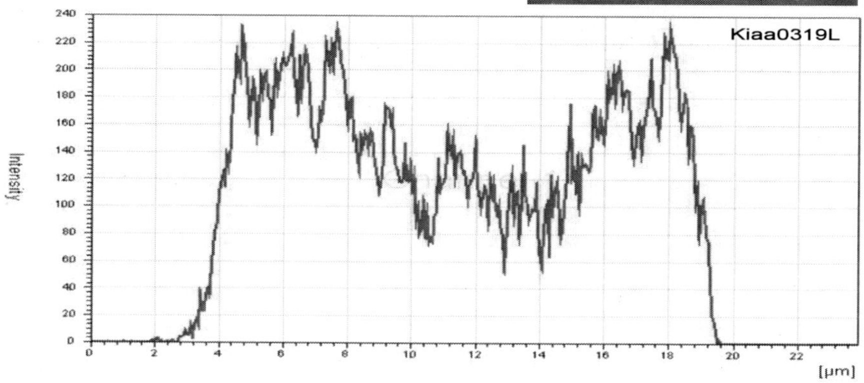

e

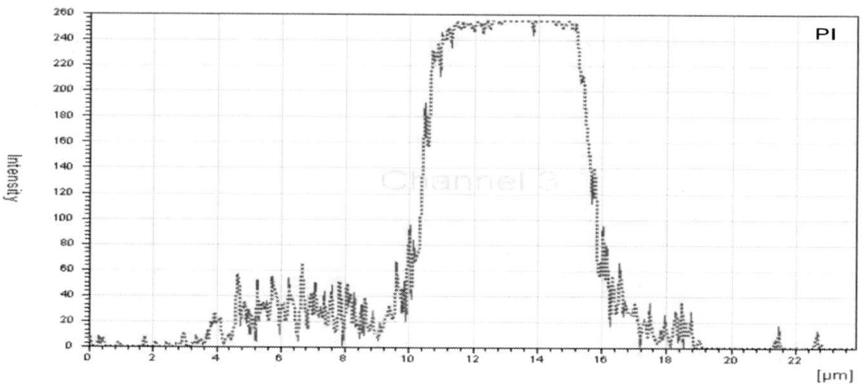

Figure 1B. Kiaa0319L proteins are predominantly expressed in cytoplasm and at the periphery of nuclei. SH-SY5Y were stained with custom rabbit polyclonal Kiaa0319L antiserum, followed by fluorescein isothiocyanate (FITC) conjugated anti-rabbit (green) secondary antibody. Cells were covered with propidium iodide, PI (10 μ g/ml) with RNase A(100 μ g/ml) in PBS to visualize the nuclei. *a. Immunofluorescence confocal analyses of celluar localization of Kiaa0319L in human neuroblastoma SH-SY5Y stained with specific anti- Kiaa0319L antibody.* (Supplementary Figure 1A): Kiaa0319L (green) is expressed endogenously in SHSY5Yin the entire cell and it is expressed strongly in the cytoplasm, neurites and weakly in the nucleus (Supplementary Figure 1B): PI staining the neuclei of SH-SY5Y (red) (Supplementary Figure 1C): bright view (Figure 1Ba): overlay view of supplementary figure1 A, B and C (Supplementary Figure 1D): inserted in (Supplementary Figure 1C) the overlay view of control staining without primary antibody. Scale bar: 10 µm. *b. Confocal analysis of a single cell of SH-SY5Y.* Confocal analysis of KL expression (green) in neurons relative to propidium idodie (PI) nucleus immunoreactivity (red). Z-stack projections through optical sections indicated by the coordinate lines are shown for a cell to the right and at the bottom of the figure. Kiaa0319L is expressed at the cell bodies and is colocalized at the surface of the PI positive nuclei whereas very little colocalized signal is found in the middle of the PI nucleus. Scale bar: 10 µm. *c Confocal analysis of Kiaa0319L expression (green) in neurons relative to PI nucleus immunoreactivity (red) represented in their intensity level of each channel.* The straight line passing through the cell showing the cross sections in analysis corresponding to the intensity curves for the indicated channels for *d. Kiaa0319L* and *e. PI nucleus staining.*

Kiaa0319L is predominantly localized in the cytoplasm, localization of Kiaa0319L is also observed in the nucleus, especially in the periphery (Figure 1B).

Characterization of Kiaa0319L Protein Was over-Expressed in Human Cell Lines

For further characterization of the Kiaa0319L protein, it was overexpressed in HEK. Having 1049 amino acid residues, the expected size of the full-length Kiaa0319L protein is 115kDa, and the expected sizes of myc- tagged and GFP-tagged proteins are 116.6kDa and 140.8kDa respectively. The overexpressed Kiaa0319L-myc protein was investigated using Western blotting of the specific antibody against the Kiaa0319L protein. Protein lysates from transiently transfected cells were separated by SDS-PAGE and subjected to Western blotting with anti-Kiaa0319L antibody. Three bands were apparent for Kiaa0319L at 190, 160 and 140kDa and all of these bands are larger than the expected size of the protein at 118kDa (Figure 2A). We used the same antibody to detect the endogenous human Kiaa0319L protein in cell lines: WRL and SH-SY5Y. Figure 2A shows bands of ~190kDa, 160kDa and 140kDa detected in both cell lysates. These results are consistent with those obtained with the over-expressed protein. (Figure 2B).

From the Western blotting results, it can be seen that the protein band sizes were larger than those expected for all these isoforms. Kiaa0319L has several N-glycosylation consensus sites, which are found in different isoforms of human and rodent Kiaa0319 and Kiaa0319L proteins (Figure 3A) so it was probable that these larger band sizes were due to glycosylation.

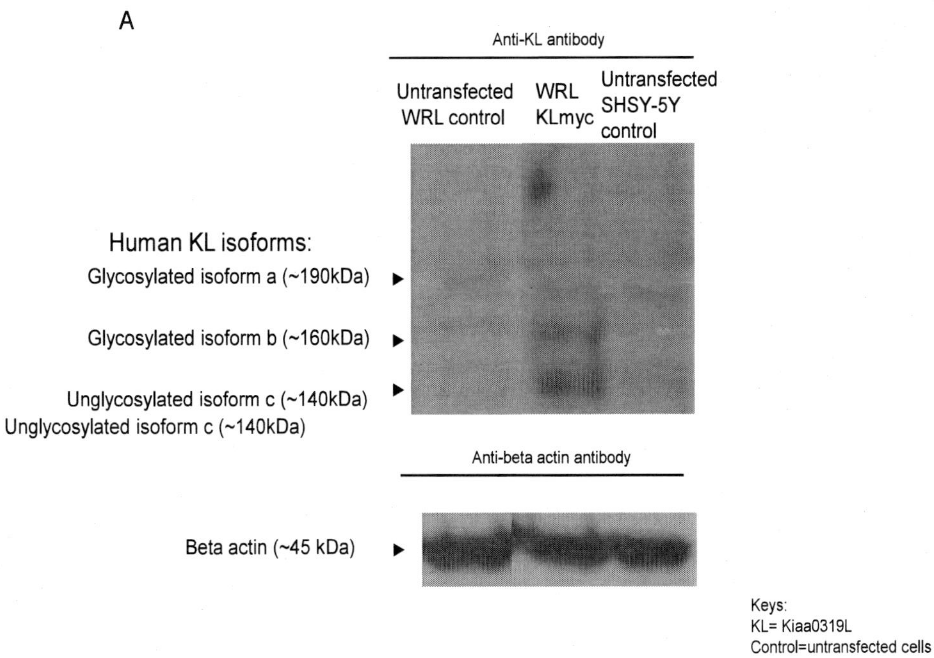

Figure 2. (Continued).

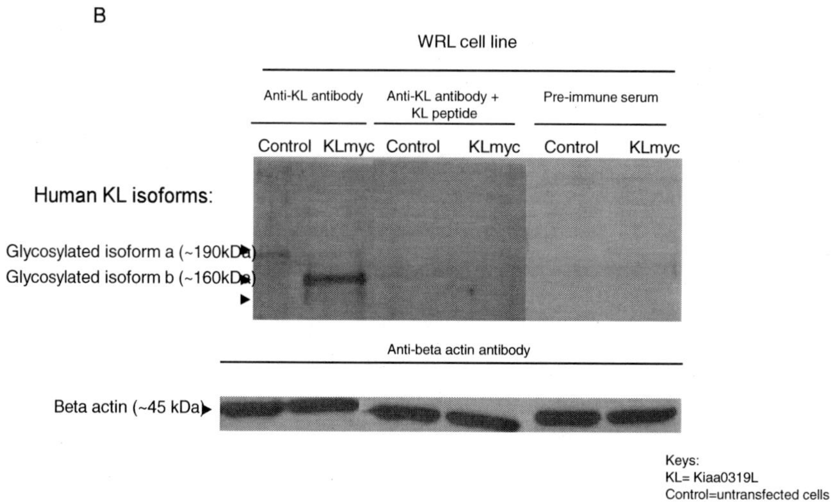

Figure 2. (A) Western blot analyses on endogenous Kiaa0319L expression in WRL and SHSY-5Y human cell lines. 3 bands of sizes about 190kDa, 160kDa and 140kDa were detected.(B) Western blot analyses testing for the specificity of rabbit polyclonal anti-Kiaa0319L antibody in human WRL-68 cell line over expressing myc-tagged Kiaa0319L protein. 3 bands of sizes about 190kDa, 160kDa and 140kDa were detected with anti-Kiaa0319L antibody in WRL overexpressing myc- tagged Kiaa0319L protein. These bands were not observed in blots treated with anti-Kiaa0319L pre-absorbed with the antigenic peptide, and in the blot overlaid with preimmuned serum. (Additional data on anti-Kiaa0319L antibody for HEK cell line is presented in [18] Figure 2).

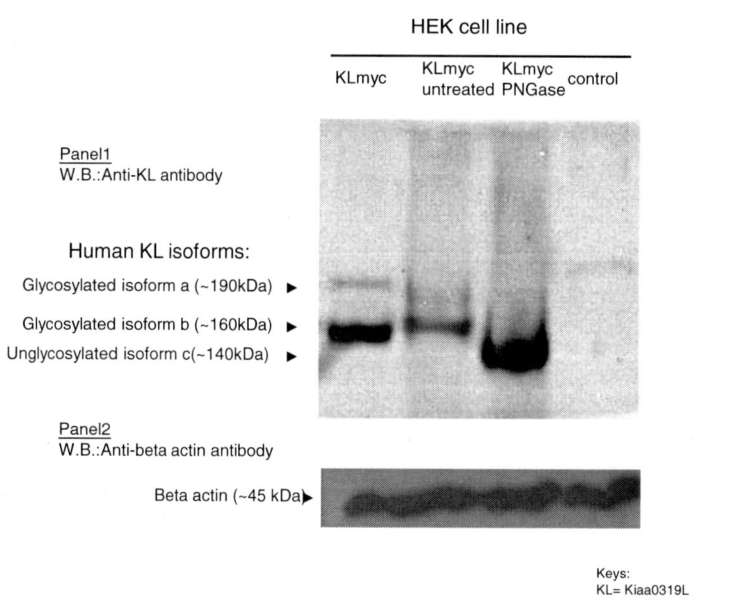

Figure 3. (A) Predicted domains and concensus N- and O- glycosylation sites of isoforms of human and rodent Kiaa0319 and Kiaa0319L proteins. (B) Western blot analyses showing N-glycosylation of Kiaa0319L. Kiaa0319L overexpressed in HEK cell lysate were treated without or with treatment of RNase F (lane2 and 3). After PNGase F treatment, the bands for 190kDa glycosylated isoform a and 160kDa glycosylated isoform b of Kiaa0319L were shifted to 160kDa and 140kDa respectively.

To test whether these increases in protein sizes were indeed due to glycosylation, the same protein lysate of the overexpressed myc- tagged Kiaa0319L were compared before and after treatment with PNGase F, an enzyme which removes nearly all N- glycosylation. Shifting in sizes were observed for both the bands from190kDa (glycosylated-isoform a) and 160kDa (glycosylated -isoform b) to 160kDa and 140kDa respectively, indicating that Kiaa0319L isoforms are N-glycosylated. Thus the 2 bands detected at 190kDa and 160kDa on expressing the full-length protein are most likely to correspond to 2 N-glylcosylated isoforms a and b (Figure 3B).

Kiaa0319L Binding to Itself

We performed co-immunoprecipitation (co-IP) experiments with protein lysate from cells co-overexpressing two types of full-length Kiaa0319L protein tagged with myc (Kiaa0319L-myc) and GFP (Kiaa0319L-GFP). Kiaa0319L-myc was detected after immunoprecipitation with anti-GFP antibody in cells co-overexpressing both tagged proteins, but not in the control cells expressing only the Kiaa0319L-myc. This result indicates that these differently tagged proteins interact with each other, and support the theory that Kiaa0319L is binding to itself (Figure 4).

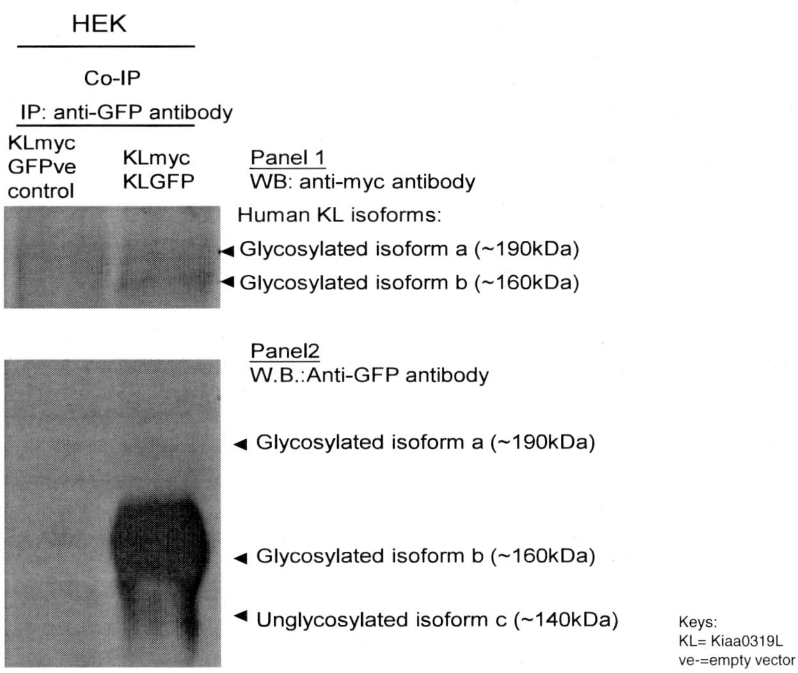

Figure 4. Interaction of human myc-tagged Kiaa0319L and GFP-tagged Kiaa0319L in human HEK cell line. Pull down of human myc- tagged Kiaa0319L with precipitated GFP tagged Kiaa0319L. Lysate of myc-tagged Kiaa0319L and GFP tagged Kiaa0319L co-overexpressing HEK cells was immunoprecipitated with (lane2) anti-GFP antibody. In the control lane (lane1), lysate of Kiaa0319L and empty GFP vector transfected HEK cells was immunoprecipitated with (lane 1) anti-GFP antibody. Immuno-complex was detected by western analysis using (Panel 1 lane 2) anti-myc antibody and (Panel 2 lane 2) anti-GFP antibody.

Reciprocal co-IP had been performed with anti-myc antibody in cells co-overexpressing both Kiaa0319L-myc and Kiaa0319L-GFP. However the results were not presented here due to a technical problem with the immunoprecipitation of Kiaa0319L-myc protein using the anti-myc antibody (which only pulls down a small fraction of the fusion protein) so that our data for Western blot staining for Kiaa0319L-GFP was not sensitive enough.

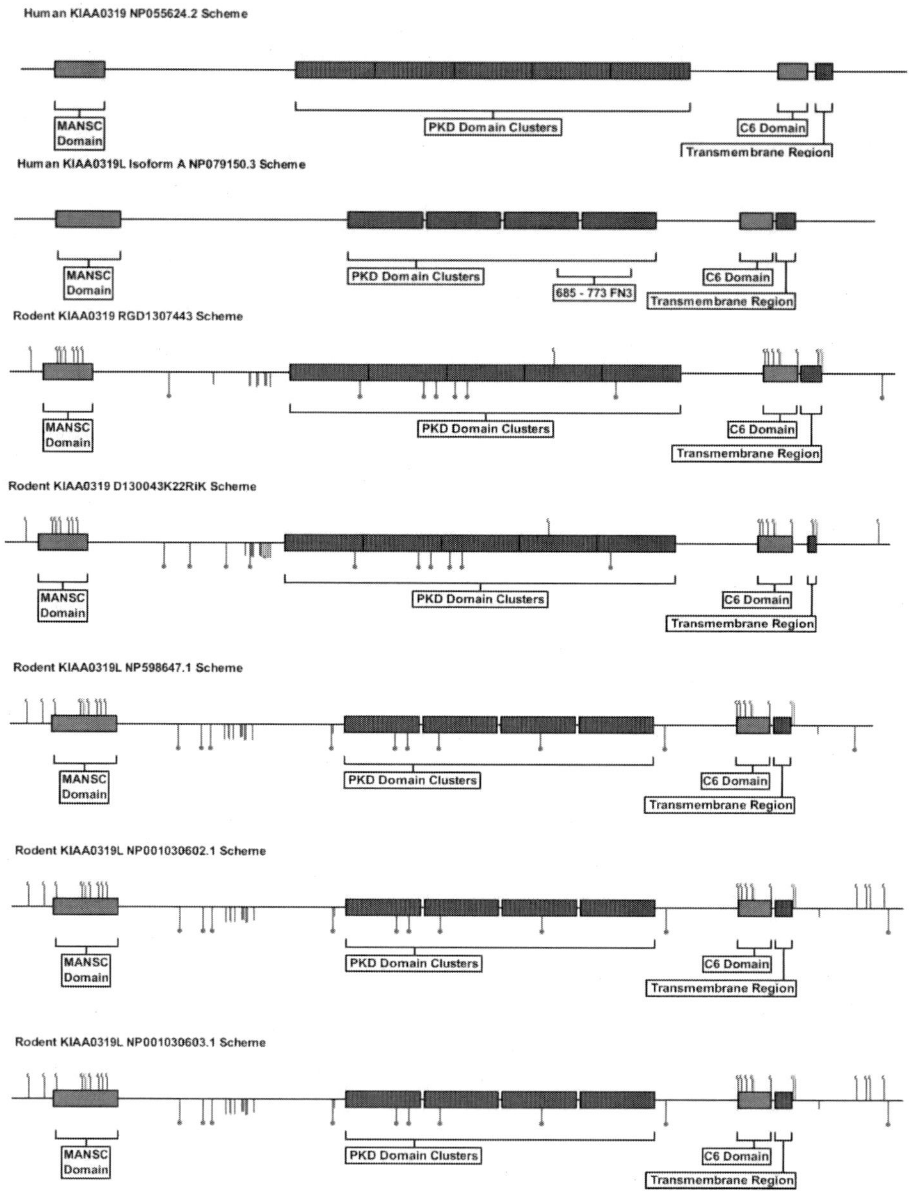

Figure 5. The human and rodent Kiaa0319 and Kiaa0319L protein isoforms. (A) Schemes of human Kiaa0319 and human Kiaa0319L encoded full-length proteins. (B) Schemes of rodent Kiaa0319 and rodentKiaa0319L encoded full-length proteins. Regions or domains detected by computational analysis are shown: MANSC domains, PKD domains, C6 domains (region of protein that includes 6 cysteine residues), and transmembrane (TM) domains. Vertical lines represent N-glycosylation consensus sites; filled circles represent O-glycosylation consensus sites.

Predicted Protein Domains on Kiaa0319 and Kiaa0319L Proteins

Human Kiaa0319 and Kiaa0319L proteins each has a MANSC domain (at its N-terminal) which is followed by several PKD domains, a C6 domain and finally a transmembrane (TM) domain before a short C-terminal tail. The PKD domains are found on protein involved in cell-cell adhesion and it is suggested to be important for adhesion between neurons and the glial fibers during neuronal migration in the Kiaa0319 protein [20]. Finding several cysteine-rich domains on the two proteins is significant: cysteine residues on TM domain of the Kiaa0319 protein are important for the dimmerization of itself while those cysteine residues on its C6 domains and MANSC domain are important for the oligomerization of itself [20] (Figure 5).

DISCUSSION

Research focusing on the genetics of dyslexia has identified several candidate genes for the molecular analysis of causal mechanisms of this disorder. Many of the candidate genes studied have been found to participate in brain development processes such as neuronal migration, axon guidance or axon growth. Specific domains and / or the homologues of these proteins could be involved in developmental processes.

KIAA0319 gene in chromosome 6p22 has been strongly associated with DD [1]. It is mainly expressed in the brain, showing a distinct expression pattern in the developing cerebral neocortex of mouse and human fetuses [6]. Interference with the expression of the homologue gene leads to impaired neuronal migration and causes apical dendrites hypertrophy [15]. The Kiaa0319 protein is a highly glycosylated, dimeric, type I plasma membrane protein and has a MANSC domain (a motif at the N-terminus with seven cysteines), PKD domains and a TM domain before its short C-terminal cytoplasmic tail [20]. This Kiaa0319 protein shares 61% homology with Kiaa0319L protein, a newly identified positional and potential functional candidate for DD [12]. Kiaa0319L is expressed abundantly in the brain [14] and it is suggested to be involved in brain development processes through its interacting partner, NgR1, an axon guidance receptor [18]. These findings suggest that these two Kiaa proteins are closely related and have complementary functions. Information on their protein homologies on their domains and protein structural properties has not yet been fully elucidated and should be a fulfilling area of research.

We show here the first experimental data for the characterization of the Kiaa0319L protein and the comparison between the two Kiaa protein homologues. Such information is important for the further investigation of their interactions and functions in the brain. The Kiaa0319L transcript is significantly expressed in both the fetal and adult brain, and is most abundantly found in the cerebellum, suggesting its physiological relevance. Our data demonstrates protein features shared by the 2 Kiaa proteins, single TM domain and several PKD domains at the N-terminus. Kiaa0319 is proposed to be involved in the interaction between neurons and glial fibers during neuronal migration. Such an interaction is believed to be mediated by the PKD domains which have been shown to form homophillic interactions and to mediate cell-cell adhesion in polycystin-1 [21].

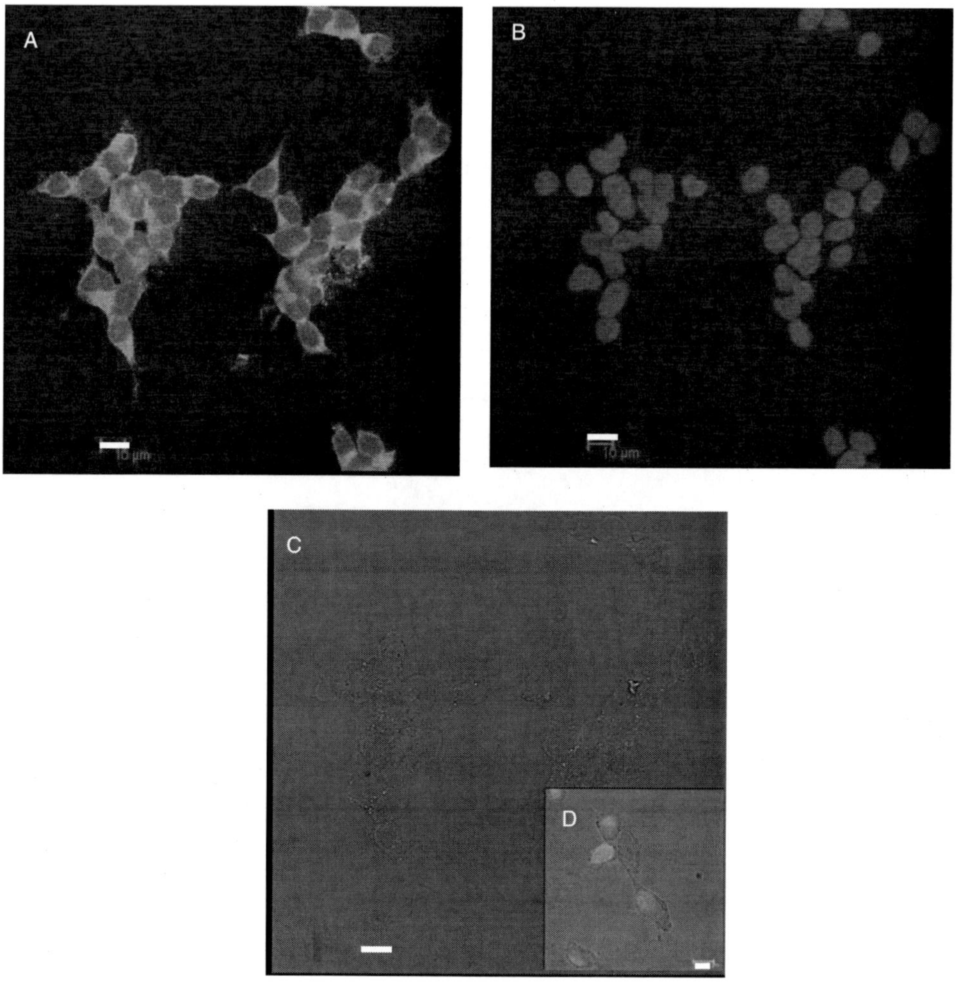

Supplementary Figure 1. Control images for immunofluorescence confocal analyses of celluar localization of Kiaa0319L in human neuroblastoma SH-SY5Y stained with specific anti- Kiaa0319L antibody. (A): Kiaa0319L (green) is expressed endogenously in SHSY5Y in the entire cell, and it is expressed strongly in the cytoplasm, neurites and weakly in the nucleus (B): PI staining the nuclei of SH-SY5Y (red) (C): bright view. (D): inserted in (C) the overlay view of control staining without primary antibody. Scale bar: 10 μm.

The distribution of the cysteine rich regions in Kiaa0319L are structurally comparable to the distribution of the MANSC domain, C6 and TM in Kiaa0319 [20]. Taken together, this study on the self-binding property of Kiaa0319L suggests that the binding of Kiaa0319L protein to itself, like Kiaa0319, is mediated by the cysteine residues in these cysteine -rich domains and disulphide bond formation between the two molecules of Kiaa0319L may be involved. The C6 regions on the two Kiaa proteins show a close similarity with the epidermal growth factor-like module (EGF domain) which forms three disulphide bridges between the 6 cysteine residues of its sequences [22, 23].

Kiaa0319L as a ligand of NgR1, axon guidance receptor, is suggested to be involved in signaling through NgR1 [24]. This signaling role of Kiaa0319L would probably be modified by the existence of the two isoforms of Kiaa0319L protein. Glycosylation has been shown to modulate interactions of receptors and ligands with themselves, thereby altering signal transduction from integrin and E-cadherin [25].

The basic characterization of Kiaa0319L protein that we report here enable us to better understand the domain structures and protein properties of the two dyslexia associated protein homologues. There are still many questions that remain unanswered; for example, why are these two Kiaa proteins binding with itself, and what are the proposed functions of the two Kiaa proteins inter-related in brain development?

CONCLUSION

Kiaa0319L as a homologue of the strong candidate encoded dyslexia-associated protein, Kiaa0319, and a dyslexia candidate gene product exhibiting similar protein properties, namely, N-glycosylation and self-binding properties. Our results will provide some hints to further investigate the close functional properties of the two Kiaa protein homologues along with their association with dyslexia and brain development.

REFERENCES

[1] S. Shaywitz, *Overcoming Dyslexia*. 2003, New York: Vintage Books.

[2] Meng, H., et al., DCDC2 is associated with reading disability and modulates neuronal development in the brain. *Proc. Natl. Acad. Sci. USA*, 2005. 102(47): p. 17053-8.

[3] Schumacher, J., et al., Strong genetic evidence of DCDC2 as a susceptibility gene for dyslexia. *Am. J. Hum. Genet*, 2006. 78(1): p. 52-62.

[4] Taipale, M., et al., A candidate gene for developmental dyslexia encodes a nuclear tetratricopeptide repeat domain protein dynamically regulated in brain. *Proc. Natl. Acad. Sci. USA*, 2003. 100(20): p. 11553-8.

[5] Dahdouh, F., et al., Further evidence for DYX1C1 as a susceptibility factor for dyslexia. *Psychiatr. Genet*, 2009. 19(2): p. 59-63.

[6] Paracchini, S., et al., The chromosome 6p22 haplotype associated with dyslexia reduces the expression of KIAA0319, a novel gene involved in neuronal migration. *Hum. Mol. Genet*, 2006. 15(10): p. 1659-66.

[7] Dennis, M.Y., et al., A common variant associated with dyslexia reduces expression of the KIAA0319 gene. *PLoS Genet*, 2009. 5(3): p. e1000436.

[8] Harold, D., et al., Further evidence that the KIAA0319 gene confers susceptibility to developmental dyslexia. *Mol. Psychiatry*, 2006. 11(12): p. 1085-91, 1061.

[9] Cope, N., et al., Strong evidence that KIAA0319 on chromosome 6p is a susceptibility gene for developmental dyslexia. *Am. J. Hum. Genet*, 2005. 76(4): p. 581-91.

[10] Hannula-Jouppi, K., et al., The axon guidance receptor gene ROBO1 is a candidate gene for developmental dyslexia. *PLoS Genet*, 2005. 1(4): p. e50.

[11] Anthoni, H., et al., A locus on 2p12 containing the co-regulated MRPL19 and C2ORF3 genes is associated to dyslexia. *Hum. Mol. Genet*, 2007. 16(6): p. 667-77.

[12] Couto, J.M., et al., The KIAA0319-like (KIAA0319L) gene on chromosome 1p34 as a candidate for reading disabilities. *J. Neurogenet*, 2008. 22(4): p. 295-313.

[13] Francks, C., et al., A 77-kilobase region of chromosome 6p22.2 is associated with dyslexia in families from the United Kingdom and from the United States. *Am. J. Hum. Genet,* 2004. 75(6): p. 1046-58.

[14] Ming-Wai Poon, M.M.Y.W.a.S.-O.C., Distribution of a novel protein encoded by the dyslexia susceptibility gene KIAA0319-like in the adult mouse brain. Under review by *Journal of Biomedical Science*.

[15] Peschansky, V.J., et al., The Effect of Variation in Expression of the Candidate Dyslexia Susceptibility Gene Homolog Kiaa0319 on Neuronal Migration and Dendritic Morphology in the Rat. *Cereb Cortex*, 2009.

[16] Threlkeld, S.W., et al., Developmental disruptions and behavioral impairments in rats following in utero RNAi of Dyx1c1. *Brain Res. Bull*, 2007. 71(5): p. 508-14.

[17] Wang, Y., et al., DYX1C1 functions in neuronal migration in developing neocortex. *Neuroscience,* 2006. 143(2): p. 515-22.

[18] Ming-Wai Poon, Wan-Hong Tsang, Sun-On Chan, Hiu-Ming Li, Ho-keung Ng, Mary Miu-Yee Waye, Dyslexia-associated Kiaa0319-Like protein interacts with axon guidance receptor Nogo Receptor 1. *Cellular and Molecular Neurobiology*, 2010 [Epub ahead of print].

[19] Velayos-Baeza, A., et al., The dyslexia-associated gene KIAA0319 encodes highly N- and O-glycosylated plasma membrane and secreted isoforms. *Hum. Mol. Genet*, 2008. 17(6): p. 859-71.

[20] Ibraghimov-Beskrovnaya, O., et al., Strong homophilic interactions of the Ig-like domains of polycystin-1, the protein product of an autosomal dominant polycystic kidney disease gene, PKD1. *Hum. Mol. Genet*, 2000. 9(11): p. 1641-9.

[21] Campbell, I.D.a.B., P, Epidermal growth factor-like modules. *Curr. Opin. Struct. Biol.*, 1993. 3(385-392).

[22] Sanderson, M.P., P.J. Dempsey, and A.J. Dunbar, Control of ErbB signaling through metalloprotease mediated ectodomain shedding of EGF-like factors. *Growth Factors*, 2006. 24(2): p. 121-36.

[23] Hasegawa, Y., et al., Promotion of axon regeneration by myelin-associated glycoprotein and Nogo through divergent signals downstream of Gi/G. *J. Neurosci*, 2004. 24(30): p. 6826-32.

[24] Zhao, Y., et al., Branched N-glycans regulate the biological functions of integrins and cadherins. *FEBS J.,* 2008. 275(9): p. 1939-48.

In: Biochemistry and Molecular Biology Coming of Age
Editor: Mary Miu Yee Waye ISBN: 978-1-63321-815-4

Chapter 5

IDENTIFICATION OF FEED FORWARD LOOPS COMPOSED OF MICRORNAS AND TRANSCRIPTION FACTORS IN ARABIDOPSIS

Nicholas Wu, Xiaozeng Yang and Lei Li*
Department of Biology, University of Virginia,
Charlottesville, VA, US

ABSTRACT

In eukaryotes, differential gene expression is regulated at many levels and by various trans-acting factors in the context of gene networks. Transcription factors (TFs) and microRNAs (miRNAs) are two classes of primary regulators for differential gene expression during organism development and function. Transcriptional gene networks mediated by TFs have been intensively mapped in several model organisms. However, relatively little has been done to examine regulatory network involving both miRNAs and TFs. This paper attempts to discover computationally the composite feed forward loops (FFLs) involving both miRNAs and TFs in the model plant *Arabidopsis thaliana* at the genome scale. 99 positional weight matrixes (PWM) that represent different DNA protein-binding motifs and 176 miRNAs were used to infer regulatory interactions at both the transcriptional and post-transcriptional levels. VisANT was then used to integrate the data and perform network motif searching. This approach generated many potential FFLs that are preferentially associated with conserved miRNAs. Further validation and study of these FFLs should provide much needed insights into the function and evolution of miRNA-mediated gene circuits in plants.

INTRODUCTION

Gene expression in eukaryotic organisms is regulated by different means including processes mediated by miRNAs at the post-transcriptional level. Derived from primary

* Corresponding Author: Lei Li, Affiliation: Department of Biology, University of Virginia, Charlottesville, VA 22904-4328, USA. Phone: 1-434-982-5481. Fax: 1-434-982-5626. Email: ll4jn@virginia.edu.

transcript containing a characteristic internal hairpin structure, mature miRNAs are small RNA molecules with a typical size of 21-24 nucleotides [1]. Similar to the biogenesis of small interference RNAs [2], miRNAs are produced by the cleavage of double-stranded miRNA hairpin structures by the Dicer or Dicer-like proteins [1]. Processed miRNA is transported to the cytoplasm where it is integrated into the RNA-induced silencing complex (RISC). The primary function of a miRNA is to base pair with the target mRNA to down-regulate gene expression either by preventing its translation or by inducing its cleavage [1].

Study of miRNA is an emerging field although an increasing body of evidence indicating that miRNA is an important regulator in a variety of biological processes. Since miRNA regulates target mRNA via a short nucleotide sequence (referred to as the miRNA binding site), multiple targeting is a characteristic of the miRNA-mediated regulatory mechanism [3, 4]. Ever since the discovery of miRNAs, this aspect of the miRNA pathway has been studied intensively. A lot of work has been done in identifying the miRNA targets in both animals and plants [3, 4]. Algorithms and experimental procedures have been developed to predict and validate the miRNA binding sites, which is critical to uncover the biological influence of each miRNA [5-7]. From these efforts, it is now clear that miRNAs function in many organism development processes and responses to environmental challenges in the context of regulatory gene circuits [1, 3].

In contrast to our knowledge of miRNA targets, relatively little has been done in examining the regulation of miRNAs themselves. In particular, the promoter regions of miRNA genes have not been well documented and understood [8]. Since the promoter region determines the regulation of the gene, study of promoter region often provides insight in the regulatory mechanism of the gene expression.

It is conceivable that miRNAs and transcription factors can potentially form many FFLs, in which one arm of the loop entails transcriptional regulation of a target gene by a transcription factor while the other arm involves a miRNA that inhibits the same target gene (Figure 1A). In fact, FFLs involving both miRNAs and transcription factors have been experimentally identified in several studies and shown to increase the complexity of gene regulatory mechanisms [9-11].

In this paper, we seek to computationally identify miRNA-transcription factor FFLs on a genome scale in the model plants Arabidopsis and rice (*Oryza sativa*). Our method is to employ bioinformatics programs to discover possible targets of the miRNAs currently annotated in Arabidopsis and rice. We also attempt to identify transcription factors that potentially bind to the miRNA promoters. By combining the results from both bioinformatics predictions, many potential FFLs are discovered. Comparison of the result from Arabidopsis to that of rice indicates that FFL is preferentially associated with conserved miRNAs.

EXPERIMENTAL SECTION

Mature miRNA sequences for Arabidopsis were obtained from miRBase (http://www.mirbase.org/). This paper employed a recently developed algorithm specifically designed for predicting plant miRNA targets [12]. In addition to structural constraints, the algorithm compares the minimum free energy (MFE) of a proposed miRNA:mRNA hybrid to that of the perfect matched duplex involving the same miRNA (MDE). A cutoff value of

MFE/MDE at 0.75 was found to have both high sensitivity and signal-to-noise ratio [12]. PlantTFBS [13] was used to determine the location of binding motif in the 1000 bp upstream region of each gene in Arabidopsis (http://www.arabidopsis.org/), including miRNA genes. These regions were then scanned for putative transcription factor binding motifs using the PWMs of 99 known binding motifs constructed in Arabidopsis [8, 13]. The Log-likelihood scoring function and the associated threshold scores for individual matrices were the same as previously described [8]. Results from PlantTFBS and miRNA targeting analyses were integrated in VisANT [14]. FFLs involving both miRNAs and TFs were searched by using a build-in function in VisANT [14].

A further restriction was applied for FFL discovery in which a set of experimentally verified miRNA target genes [15] was used to develop the link between the miRNA and the target genes in the FFL searching process.

The same analyses were also performed on the rice genome (http://rice.plantbiology.msu.edu/) with the miRNA sequences obtained from RNABase. Homologous genes between rice and Arabidopsis were identified by using Blastp to align the annotated protein sequences in rice with those in Arabidopsis with the e-value cut-off set at $1e^{-10}$. Homologous FFLs were identified by comparing the results in rice and Arabidopsis. A homologous FFL was defined as a unit that includes an identical miRNA gene, an identical DNA binding motif and a homologous protein between rice and Arabidopsis.

RESULT AND DISCUSSION

We are interested in three-gene FFLs composed of one input TF and one input miRNA regulated by the TF, which jointly regulate a target gene (Figure 1A). To provide a genome wide analysis of such FFLs, we developed the following strategy. First, we used an algorithm specifically tailed from the program RNAhybrid to predict miRNA target genes [12]. Next, we utilized PlantTFBS [13], a Java script based program, to predict the TF binding site in the putative miRNA promoter regions as well as the promoters of the target genes based on 99 known binding motifs in Arabidopsis constructed by Megraw et al. [8]. Finally, results from both bioinformatics predictions were integrated using VisANT [14] to identify potential feed forward loops. In total, we found 2921 FFLs in Arabidopsis. If only experimental verified miRNA targets were used, 1380 loops were obtained. It should be noted that our method did not distinguish between members of the same miRNA or TF families. Therefore, the number of identified FFLs was probably over-estimated. Nevertheless, we believe the relative frequency of FFLs in our dataset is still informative about the context in which miRNAs function and regulation are carried out.

Out of the 176 annotated miRNAs in Arabidopsis, 113 (64.2%) were found to be associated with at least one FFL. To facilitate further analysis of the FFLs, we divided the miRNA into three groups. Group I includes 21 miRNA families (156/157, 160, 162, 164, 166-169, 171, 172, 319, 390, 393-399 and 408) that are universal among diverse plants [16]. Group II includes 10 families that are conserved in at least two plant species. Group III contains approximately 70 families that are only found in Arabidopsis thus far. We then calculated the relative frequency of FFLs associated with miRNAs in these three different groups. We found that Group I (highly conserved) miRNAs on average are associated with

29.0 FFLs, Group II (intermediately conserved) miRNAs on average are associated with 13.9 FFLs, whereas the non-conserved Group III miRNAs are only associated with 3.4 FFLs on average (Figure 1B). These results indicate that miRNAs present in FFLs are highly enriched with the conserved ones.

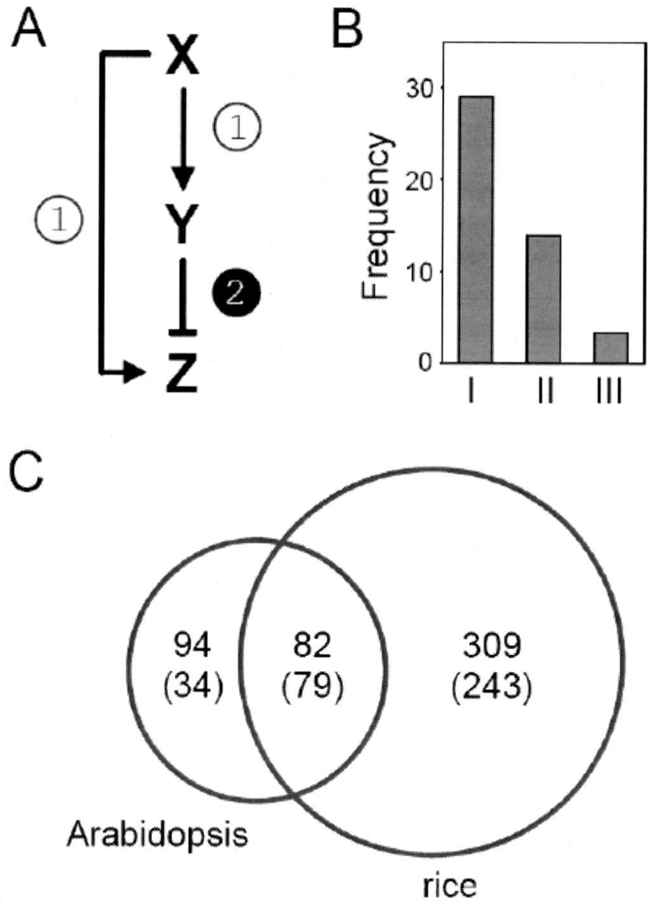

Figure 1. Analyses of three-gene FFLs involving both miRNAs and TFs. (A) The structure of the FFL is illustrated in which gene X represents an input TF, Y represents a miRNA and Z represents the common target gene. 1, PlantTFBS was used to predict the existence of specific DNA motif in both miRNA and gene. 2, RNAhybrid was used to predict miRNA targeting. (B) Relative frequency of FFL was calculated as the average number of FFL per miRNA for Group I, II and III miRNAs. (C) Venn diagram showing the number of miRNAs conserved and not conserved between Arabidopsis and rice. The numbers in the parentheses represent the miRNAs associated with FFLs.

For a cross-species comparative analysis, we searched the rice genome for FFLs as well. We define homologous FFLs as the three-gene patterns in which all three are orthologous between Arabidopsis and rice. Through this analysis, we found that there are 636 FFLs in Arabidopsis that are homologous to 975 in rice. For those 636 loops in Arabidopsis, 615 are involving with experimentally verified miRNA targets. We found 82 miRNA are conserved between Arabidopsis and rice while 94 are unique to Arabidopsis. Interestingly, 79 (96%) of the conserved ones are involved in FFL while only 36% of the unique ones are (Figure 1C).

These results indicate that miRNAs conserved between rice and Arabidopsis are most likely to be present in the FFLs in both plant species.

It has been argued that new miRNAs continue to appear in evolution. In addition, those miRNAs that provide selective advantages would be retained and hence become conserved in multiple species [1, 17]. We found that conserved miRNAs are preferentially associated with FFLs. Thus, it implies that incorporation into a FFL could be one of the mechanisms for the retention of newly evolved miRNAs in plants. Involvement in the FFLs probably has given these miRNAs the opportunity to participate in specific gene regulatory mechanisms and thus an evolutionary advantage. Conversely, the low participation rate of non-conserved miRNAs in the FFL mechanism could probably be due to lack of sufficient time for the evolution of newly evolved miRNAs [1, 17].

In this paper, the primary goal of our approach is to computationally identify all the possible transcriptional and post-transcriptional interactions in order to generate hypothetic FFLs with relative high confidence level. Since our knowledge of these interactions is still rather limited, it is inevitable that the computationally inferred interactions contain both false positives and false negatives. For example, the PWM might represent a class of TFs instead of a specific TF. Thus, the promoters that share the same motif might not be sharing the same regulatory TF. However, in addition to the evolutionary considerations, our data could be used in conjugation with the knowledge of specific biological pathways or experimental systems to further study the miRNA-mediated gene circuits. Since in specific metabolic pathways or homeostasis systems, only certain TFs or miRNAs will be involved. Prior knowledge of the pathway can be used to guide the selection of specific FFLs in order to narrow down the candidates. In this fashion, specific hypothetical loops can be selected and experimentally tested.

As an example, we discuss a particular FFL identified in our analysis. This FFL involves miR399 and the TFs in the WRKY family that recognize the W-box. By searching the literature, we found that WRKY75 is a TF that responds to phosphate deficiency in Arabidopsis, which is known to bind to the W-box DNA motif [18]. The promoter region of miR399, which has been shown to respond to phosphate deficiency, possesses a W-box motif. In addition, the gene AT2G33770, which encodes an ubiquitin-conjugated enzyme, has been experimentally verified as being the target of miR399 [19] and also has a W-box in its promoter. As a result, we can now propose a distinct FFL consisting of WRKY75, miR399 and the common target gene AT2G33770. It demonstrates the usefulness of the computational approach in generating highly confident hypothetical FFLs. It is suggested that the interactions in this loop should be validated in the future to enhance our knowledge of the biological responses phosphate deficiency in Arabidopsis as well as other plants that contain the same FFL.

CONCLUSION

In this study, we describe a computational approach to identify hypothetical FFLs consisting of both miRNAs and TFs in Arabidopsis on a genome-wide scale. We identified a large number of FFL candidates that involve over half of the known miRNAs. Although we could not rule out the false positive and the false negative results due to the lack of large scale experimental data, our results provide an important insight into miRNA evolution. Future examination in other model organisms should allow us to verify whether conserved miRNAs tend to participate in FFLs more often than the non-conserved miRNAs do. In addition, our data can be used in conjugation with prior knowledge of specific pathways to identify highly confident hypothetical FFLs that can be experimentally tested to accelerate our understanding of the miRNA-mediated gene circuits.

REFERENCES

[1] Voinnet, O. Origin, biogenesis, and activity of plant microRNAs. *Cell.* 2009; 136: 669-687.

[2] Jaronczyk, K; Carmichael, J; Hobman, T. Exploring the functions of RNA interference pathway proteins: some functions are more RISC than others? *Biochem. J.* 2005; 387: 561–71.

[3] Stark, A; Brennecke, J; Bushati, N; Russell, RB; Cohen, SM. Animal MicroRNAs confer robustness to gene expression and have a significant impact on 3'UTR evolution. *Cell.* 2005; 123: 1133–46.

[4] Bartel, DP. MicroRNAs: genomics, biogenesis, mechanism, and function. *Cell.* 2004; 116: 281-297.

[5] Wang, X; Naqa, IM. Prediction of both conserved and nonconserved microRNA targets in animals. *Bioinformatics.* 2008; 24: 325-332.

[6] Yousef, M; Jung, S; Kossenkov, AV; Showe, LC; Showe, MK. Naïve Bayes for microRNA target predictions—machine learning for microRNA targets. *Bioinformatics.* 2007; 23: 2987-2992.

[7] Kruger, J; Rehmesmeier, M. RNAhybrid: microRNA target prediction easy, fast and flexible. *Nucleic Acids Research.* 2006; 34: 451-454.

[8] Megraw, M; Baev, V; Rusinov, V; Jensen, ST; Kalantidis, K; Hatzigeorgiou, AG. MicroRNA promoter element discovery in Arabidopsis. *RNA.* 2006; 12: 1612-1619.

[9] Hornstein, E; Shomron, N. Canalization of development by microRNAs. *Nat. Genet.* 2006; 38: S20-S24.

[10] Cohen, EE; Rosner, MR. MicroRNA-regulated feed forward loop network. *Cell Cycle.* 2009; 8: 2477-2478.

[11] Re, A; Cora, D; Taverna, D; Caselle, M. Genome-wide survey of micro-RNA-transcription factor feed-forward regulatory circuits in human. *Mol. BioSys.* 2009; 5: 854-867.

[12] Alves-Junior, L; Niemeier, S; Hauenschild, A; Rehmsmeier, M; Merkle, T. Comprehensive prediction of novel microRNA targets in *Arabidopsis thaliana. Nucleic Acids Research.* 2009; 37: 4010-4021.

[13] Megraw, M; Hatzigeorgiou, AG. MicroRNA promoter analysis. *Methods Mol. Biol.* 2010; 592: 149-161.

[14] Hu, Z; Mellor, J; Snitkin, SE; Delisi, C. VisANT: an integrative framework for networks in systems biology. *Bioinformatics.* 2008; 9: 317-325.

[15] German, MA; Pillay, M; Jeong, DH; Hetawal, A; Luo, S; Janardhanan, et al. Global identification of microRNA-target RNA pairs by parallel analysis of RNA ends. *Nat. Biotechnol.* 2008; 26: 941-946.

[16] Axtell, MJ; Bowman, JL. Evolution of plant microRNAs and their targets. *Trends Plant Sci.* 2008; 13: 343-349.

[17] Chen, K; Rajewsky, N. The evolution of gene regulation by transcription factors and microRNAs. *Nat. Rev. Genet.* 2007; 8: 93-103.

[18] Bao, J; Zhang, S; Lu, W; Guo, C; Gu, J; Xiao, K. Transcriptional responses and regulations to deficient phosphorus in plants. *Front. Agric. China.* 2009; 3: 47-54.

[19] Bari, R; Datt PB; Stitt, M; Scheible, WR. PHO2. microRNA399, and PHR1 define a phosphate-signaling pathway in plants. *Plant Phsiol.* 2006; 148: 3-5.

In: Biochemistry and Molecular Biology Coming of Age
Editor: Mary Miu Yee Waye
ISBN: 978-1-63321-815-4

Chapter 6

PROTEOMIC PROFILING OF 6-HYDROXYDOPAMINE-TREATED ZEBRAFISH USING iTRAQ-2D LC-MS/MS ANALYSIS

***Benson U. Wang Lei*[1,†], *Nan Yu*[2,†], *Zai-Jun Zhang*[1,3,†], *Edward Lau*[2,†], *S. O. Siu*[1, 2], *Maggie P. Y. Lam*[2], *Guo-Hui Li*[2], *Ivan Keung Chu*[2] *and Simon Ming-Yuen Lee*[1,*]**

[†] These authors contributed equally to this work.

[1]State Key Laboratory of Quality Research in Chinese Medicine, Institute of Chinese Medical Sciences, University of Macau, Taipa, Macao SAR, China

[2]Department of Chemistry, The University of Hong Kong, Hong Kong SAR, China

[3]Institute of New Drug Research and Guangdong Province Key Laboratory of Pharmacodynamic Constituents of Traditional Chinese Medicine and New Drug Research, Jinan University College of Pharmacy, Guangzhou, China

ABSTRACT

6-Hydroxydopamine (6-OHDA), one of the most commonly used dopaminergic (DA) neurotoxins used as a model for Parkinson's disease (PD) *in vitro* and *in vivo*, is likely to damage neurons through the generation of reactive oxygen species (ROS) such as the superoxide radical in mitochondria; however, details of the molecular events that occur in such a model are not available. In an attempt to identify the proteins involved in this process, we applied an iTRAQ-2D LC-MS/MS-based proteomic approach to analyze the protein profile of the 6-OHDA-induced PD zebrafish (*Danio rerio*) model. Zebrafish were exposed to 6-OHDA to selectively kill DA neurons. DA depletion was determined by whole mount immunostaining with antibody against tyrosine hydroxylase and was further validated by the detection of tyrosine hydroxylase (*th*) mRNA expression using quantitative real-time PCR. The change of zebrafish swimming behavior as a

* Corresponding author: Institute of Chinese Medical Sciences, University of Macau, Av. Padre Toma´s Pereira S.J., Taipa, Macao SAR, China; Tel: +853 8397 4695, Fax: +853 2884 1358; E-mail address: simonlee@ umac.mo (Simon M.Y. Lee).

consequence of DA neuronal damage was monitored. In contrast to vehicle-treated healthy zebrafish, treatment with 6-OHDA induced DA neuronal loss, down-regulated *th* mRNA expression and significantly altered the swimming behavior of zebrafish. Our iTRAQ-2D LC-MS/MS analysis identified a total of 1,709 unique proteins from zebrafish lysates and bioinformatic analysis of the proteomic profile suggested the potential involvement of both TNF-α/NF-κB and oxidative phosphorylation pathways in 6-OHDA-induced neurodegeneration in zebrafish. These data emphasize the need for further research at the global protein level to unravel the underlying mechanism of 6-OHDA-induced neurotoxicity and the roles of the corresponding targets.

INTRODUCTION

Parkinson's disease (PD) is a degenerative disorder of the central nervous system (CNS) that impairs motor skills, cognitive processes and other functions. PD is characterized by a severe loss of dopaminergic (DA) neurons resulting in a range of motor deficits. DA deficiency is produced in the brain of an experimental animal by central or systemic administration of specific DA neurotoxins; the most commonly used are 6-hydroxydopamine (6-OHDA) and 1-methyl-4-phenyl-1,2,3,6-tetrahydropyridine (MPTP) [1]. 6-OHDA is a hydroxylated analog of the natural neurotransmitter dopamine [2]. Its induced toxicity is relatively selective for catecholaminergic neurons, resulting from a preferential uptake of 6-OHDA by dopamine and noradrenergic transporter molecules [3].

Accumulating evidence suggests that oxidative stress plays a pivotal role in DA neurodegeneration. It has been reported that 6-OHDA-induced neuron degeneration involves the processing of hydrogen peroxidase and hydroxyl radicals in the presence of iron [4]. Furthermore, it has been shown that treatment with 6-OHDA reduces striatal glutathione (GSH) and superoxide dismutase (SOD) enzyme activity [5] and increases the levels of malondialdehyde [6]. 6-OHDA also appears to be toxic to mitochondrial complex I [7] and leads to the formation of superoxide free radicals [8]. However, the mechanism underlying 6-OHDA-induced neuronal damage is not fully understood.

Proteomics is a powerful methodology that can be used to investigate how protein expression is affected during pathogenesis, providing a complement to the information obtained by functional genomics [9]. Earlier proteomic studies used mainly integrated technologies, including separation of proteins by two-dimensional polyacrylamide gel electrophoresis (2D-PAGE) and identification of proteins by matrix-assisted laser desorption/ionization time-of-flight mass spectrometry (MALDI-TOF MS). Together, these techniques logically open the door to a proteomics study of the 6-OHDA-induced PD model.

Recent advances in the field of proteomics have made it possible to profile comprehensive protein expression levels and to quantify the proteome under different conditions with various quantitative proteomics techniques [10]. The use of zebrafish (*Danio rerio*) embryos for proteomics research allows direct incorporation of evidence at both the organism and protein level [11]. Quantitative proteomics data from zebrafish can serve as a basis for identifying unknown molecular pathways for neuronal damage of 6-OHDA that could readily be coupled to visual observations.

In order to investigate global protein expression changes *in vivo*, we applied quantitative mass spectrometric proteomics with isobaric tags for relative quantification (iTRAQ), a unique *in vitro* isotopic labeling strategy for quantifying up to 8-plex samples simultaneously

[12] and to acquire protein expression ratios in protein lysates from control and 6-OHDA-treated zebrafish. Although components of known neuronal damage pathways will likely provide most direct information about the observed changes, one of the advantages of comparative proteomics lies in its neutrality to *a priori* knowledge and targets. The observation-driven results could lead to insights into previously unsuspected targets.

Several factors make the zebrafish embryo platform an attractive choice for the detection of protein variation *in vivo* with proteomics [13, 14]. Firstly, the zebrafish is an established vertebrate model organism and has been used to study development of the nervous system [15]. Zebrafish and human are remarkably similar in biochemistry and physiology, and show high degrees of conservation in neurogenesis pathways. Secondly, the high level of fertility, short generation time and well defined external embryonic developments make the zebrafish highly amenable for studying developmental changes [16]. Lastly, the visual transparency of the zebrafish embryos and the availability of whole-mount immunohistochemistry without sectioning allow direct observation of neuronal cells by microscopy. With that in mind, the major goal of the present study was to understand, at the protein level, the molecular mechanism underlying the neurodegenerative action of 6-OHDA in a viable zebrafish model *in vivo*.

METHODS

Zebrafish Maintenance and Drug Treatment

AB strain wild type zebrafish were used for this study. Embryos were collected after natural spawning, staged according to standard criteria, and raised synchronously at 28.5°C in embryo medium (13.7 mM NaCl, 540 μM KCl, 25 μM Na_2HPO_4, 44 μM KH_2PO_4, 300 μM $CaCl_2$, 100 μM $MgSO_4$, 420 μM $NaHCO_3$, pH 7.4). No additional maintenance was required because embryos receive nourishment from the attached yolk ball.

Ethical approval for these animal experiments was granted by the Animal Research Ethics Committee, University of Macau.

For drug treatment, zebrafish were dechorionated at 1 day post-fertilization (dpf), and exposed to 250 μM 6-OHDA in embryo medium for 48 h. After treatment, fish were collected for whole mount immunostaining and preparation of total RNA or proteins. In the swimming behavior test, 4 dpf zebrafish larvae were exposed to 250 μM 6-OHDA for 72 h and swimming behavior was monitored at 7 dpf. Untreated, wild type zebrafish were used as the negative control.

Anti-Tyrosine Hydroxylase Whole Mount Immunostaining

Zebrafish were fixed in 4% (v/v) paraformaldehyde in PBS for 5 h, rinsed, and stored at –20°C in 100% EtOH. Whole mount immunostaining was done by standard methods [17]. Briefly, fixed samples were blocked (2% (v/v) lamb serum and 0.1% (w/v) BSA in PBS, at room temperature for 1 h. A mouse monoclonal anti-tyrosine hydroxylase antibody (1:200 diluted in blocking buffer, MAB318, Millipore) was used as the primary antibody and was incubated with the sample at 4°C overnight. The next day, samples were washed six times with PBST (30 min each wash), followed by incubation with secondary antibody according to the instructions provided with the Vectastain ABC kit (Vector Laboratories, Inc.). After staining, zebrafish were flat-mounted with 3.5% (w/v) methylcellulose and photographed.

Total RNA Extraction and Reverse Transcription and Real-Time Quantitative PCR (qPCR)

Total RNA was extracted from 40 zebrafish embryos of each treatment group using the RNeasy Mini Kit (Qiagen, USA) according to the manufacturer's instructions. RNA was reverse transcribed to single-stranded cDNA using the SuperScript™ III First-Strand Synthesis System for RT-PCR (Invitrogen™, USA), followed by real-time PCR using the TaqMan® Universal PCR Master Mix and TaqMan® gene expression assay primer for zebrafish tyrosine hydroxylase gene (*th,* assay ID Dr03437803_g1 from Applied Biosystems) in the ABI 7500 Real-Time PCR System (Applied Biosystems). The level of expression of the *th* gene is presented as relative fold change ($\log_2$) calculated using the comparative C_t method as described [18]. β-Actin was used as the internal reference; the zebrafish *bactin1* primers were synthesized by Applied Biosystems:

forward 5′-CAAGATTCCATACCCAGGAAGGA-3′
reverse 5′-CAAGATTCCATACCCAGGAAGGA-3′

Swimming Behavior

At 3 dpf, a single larva was pipetted into each well of a 12-well plate containing 2 mL of embryo medium then exposed to 6-OHDA (final concentration 250 μM) by pipetting 100 mM stock solution into each well (6 animals/group). After 72 h, swimming behavior was monitored using an automated video tracking system; the 12-well plate and camera were housed inside a Zebrabox (ViewPoint, LifeSciences). The swimming pattern of each fish was recorded for 10 min. The total distance moved (in cm) was defined as the distance that the fish moved during one session (10 min).

Total Protein Extraction

200 one dpf control and 6-OHDA treated zebrafish embryos were transferred to two separate 1.5 mL microfuge tubes and de-yolked using calcium-free deyolking buffer as described [19]. Then 30 μl of dissolution buffer (iTRAQ kit) containing 0.3 μl of protease inhibitor PMSF and 0.3 μl of phosphatase inhibitor Cocktail2 (Sigma, Saint Loius, MO, USA) was added, followed by sonication in an ice bath for 5 min with each 1 s sonication burst followed by a 3 s pause.

The lysates were centrifuged at 14,000*g* for 25min at 4°C and then the supernatant was recovered. The concentration of the lysate was determined using a Bio-Rad protein determination kit (Bio-Rad; Hercules, CA) according to the manufacturer's instructions using bovine serum albumin (BSA) as the calibration standard. The proteins were stored at –80°C.

Trypsin Digestion and Labeling

Each 100 μg protein sample was reduced to the volume about 20 μL by speedvac. After denatured by 1ul 1% SDS, then add 2ul 50mM TCEP and incubate at 60°C for 1h followed by adding 1ul 200mM MMTS and incubate at room temperature for 10 minutes by following the iTRAQ Reagents-8 plex kit protocol provided with the equipment.

Then the proteins were digested with W (protein): W (trypsin) = 33: 1 sequencing-grade trypsin (Promega, Madison, WI) at 37°C overnight in the thermomixer and then labeled with the appropriate tag at room temperature for 2 h by following the kit instructions.

Strong Cation-Exchange (SCX) Chromatography

The combined iTRAQ samples were diluted into ~4 mL of 10mM H_3PO_4 15% (v/v) acetonitrile (ACN) pH 3 before fractionation by SCX with a Perkin Elmer series 200 LC system equipped with a PolyLC PolySULFOETHYL A™ column (100 mm × 2.1 mm, 5 μm particles, 300 Å pore size).

The elution was performed with a 60 min gradient from 0% B to 100%B at the flow rate of 0.2 ml/min. Solvent A: 10mM H3PO4/KH2PO4, 20%ACN, PH=3; Solvent B: 10mM H3PO4/KH2PO4, 350mM KCl, 20%ACN, PH=3. The eluent was monitored using an Agilent 1100 diode-array detector module (Agilent Technologies; Santa Clara, CA) set to measure UV absorption at 220 nm wavelength with a bandwidth of 6 nm. Fractions were collected every 1 min from the time UV absorption was observed to increase after the initial flow-through until the absorption was judged to have reverted to near-baseline. A total of 30 fractions with absorption at 220/6 > 1000 units were collected and dry by speedvac at low temperature.

Capillary LC-MS/MS

Each SCX fraction was reconstituted in 20 μl of solvent A for individual analysis. A 5 μl sample of resolved fractions was injected into a self-packed RP capillary column (I.D 150 μm, length 10 cm). Agilent 1100 series capillary pump equipped with a one well-plate auto-sampler (Agilent Technologies,Wilmington, DE, USA) coupled with QSTAR XL quadrupole/time-of-flight hybridmass spectrometer (AB SCIEX, Foster City, CA, USA) controlled by Analyst 1.4.2 software was employed for LC-MS/MS at a flow rate of 1.5 μl/min with a linear gradient of 5% B to 35% B over 120 min. Mobile phase A is 98% (v/v) water and 2% ACN with 0.5% (v/v) formic acid and mobile phase B is 98% ACN and 2% water with 0.5% formic acid.

The optimized MS parameters were as follows: DP, 60 V; FP, 220 V;DP2, 25 V; CAD, 5; GS1, 0 psi;GS2, 0 psi; CUR, 20; nanospray voltage, 3200 V. Information dependent acquisition (IDA) was applied in which every 1-s survey scan in the range of mass-to-charge (*m/z*) 400–1500 was followed by up to four 1-s MS/MS scans of the four most intense peaks in the range *m/z* 100–1500 with more than 20 counts and charge states between +2 and +4. The dynamic exclusion of acquired ions was performed for 90 s to avoid redundant product ion scans. Enhance-all ion settings were used for MS/MS.

Analysis of Tandem Mass Spectrometry (MS/MS) Data

The acquired MS/MS data were analyzed directly using the Paragon algorithm in ProteinPilot 3.0 software (Applied Biosystems, Foster City, CA). The MS/MS spectra were matched against theoretical spectra generated from sequences in the International Protein Index (IPI) release 3.56 *Danio rerio* (43,489 entries) protein database (http://www.uniprot.com).

For the profiling experiment, trypsin was the enzyme used, iodoacetamide was chosen as the cysteine alkylation and thorough identification with biological modification settings, which contain a number of built-in modifications, were used; for the quantification experiments, iTRAQ 8-plex quantification was selected, with trypsin as the enzyme and *ethyl methanethiosulfonate* (MMTS) as cysteine alkylation in a thorough quantification search with biological modifications.

Precursor mass accuracy and product ion mass accuracy were pre-set by the software according to the instrument option chosen (QSTAR ESI).

The peptides identified with the Paragon algorithm were grouped into minimal non-redundant protein sets by the ProGroup algorithm of the software. A minimal unused ProtScore of 1.3 was used for protein identifications to be considered. The unused ProtScore was the sum of individual peptide scores based on identification confidence that was not assigned to higher-ranked protein sets, and it served to minimize redundancy of protein identification based on common peptide sequences.

The peptide score was based on statistical confidence of sequence matching according to the software; a protein of 99% confidence was translated into a score of 2.0 and 95% confidence was scored 1.3.

In the complex sample analysis experiment, protein sets containing two or more peptides passing the 95% confidence threshold were accepted automatically, whereas other protein sets were manually confirmed. Only peptides that were unique to the particular protein hit and with a signal-to-noise ratio greater than 9 contribute to the protein quantification according to Proteinpilot algorithm. The statistical confidence *p*-values of quantification information represented a significant deviation from unity in expression levels and were calculated by Paragon according to the default settings of the software.

Functional Analysis

The sequences of the quantified proteins were fetched from the IPI website using the dbfetch web tool. Blast2GO V.2.3.6 (http://www.blast2go.org) was used to acquire Gene

Ontology (GO) terms for the protein sequences or their close homologs. In the BLAST step, NCBI non-redundant was chosen as the database, a low-complexity filter and a minimum expected value of 1.0^{-10} was used and up to 20 BLAST results were fetched for each sequence; in the annotation step, *E*-value-Hit-Filter was set to1.0^{-6}, the annotation cutoff was 55 and the GO weight was 5 (software default settings).

Each NCBI non-redundant homolog had a minimum expected value of 1.0^{-9} or below as well as a sim mean of at least 45.6%. Figure 1 shows the outline of the proteomics study.

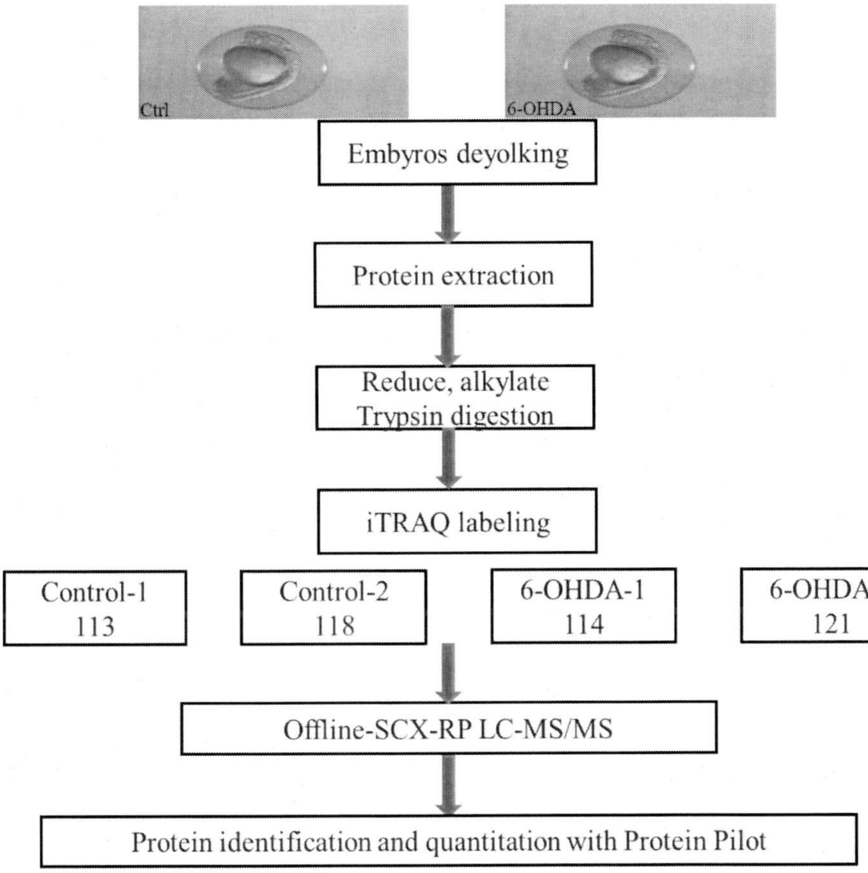

Figure 1. The experimental strategy for iTRAQ based quantitative proteomics study.

Statistical Analysis

Statistical analysis was done with GraphPad Prism 5.0 software (GraphPad Software Inc., San Diego, CA, USA). One-way ANOVA was used followed by Dunnett's test for significant difference analysis. Each experiment was done in triplicate and all error bars indicate mean ± SEM.

RESULTS

Effects of 6-OHDA on Neuronal Damage in Zebrafish

In a pilot study, zebrafish embryos at 1 dpf were exposed to serial dilutions of 6-OHDA (range 62.5–1000 μM) for 2 days (data not shown). The lowest effective concentration of 6-OHDA that can induce DA neuron loss with no systematic toxicity is 250 μM. Therefore, 250 μM of 6-OHDA was used for all the experiments in the present study. To assess the effects of 6-OHDA on DA neurons, we carried out immunostaining with anti-tyrosine hydroxylase (TH) antibody. Although anti-TH antibody stains several types of catecholaminergic neurons, it has been demonstrated that all TH-positive neurons in the zebrafish diencephalon (hypothalamus, posterior tuberculum, ventral thalamus and pretectum) are DA neurons [20] [21]. As shown in Figure 2A, populations of DA neurons were observed in the diencephalon of both control and 6-OHDA-treated zebrafish. However, following treatment with 250 μM of 6-OHDA, the number of DA neurons of the disease group was comparably much less than that in the control. Furthermore, the total number of TH-positive DA neurons was reduced significantly to 46.4 ± 17.5% in 6-OHDA-treated zebrafish ($p < 0.001$) (Figure 2B). The decrease in TH-positive DA neuron population in the zebrafish model reflected the neuronal damage induced by 6-OHDA.

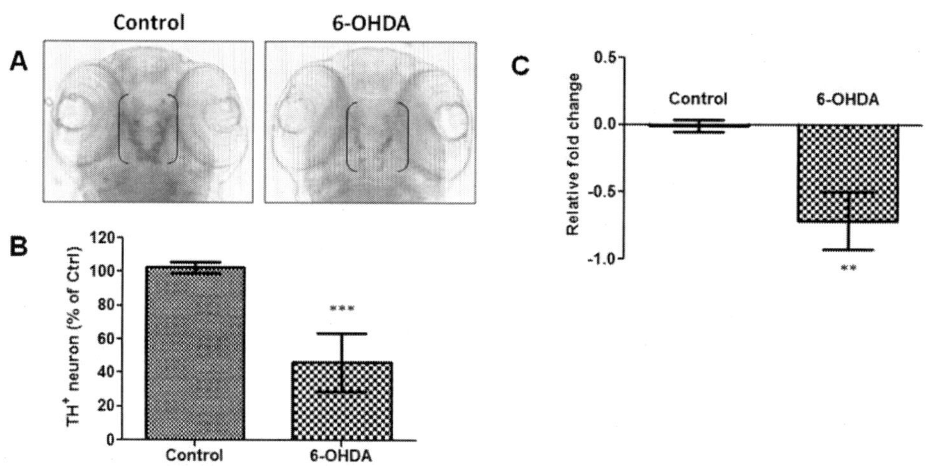

Figure 2. The 6-OHDA-induced neuronal damage in zebrafish at 72 hpf. (A) Representative picture of anti-TH whole mount immunostaining. TH+ neurons in the diencephalic region are indicated by brackets, dorsal view. (B) Count of TH+ neuron; *** $p < 0.001$. (C) Relative fold change of th gene expression as compared to that of the control: 6-OHDA down-regulated th gene expression; ** $p < 0.01$.

Tyrosine hydroxylase plays a specific and important physiological role in DA neurons. As tyrosine hydroxylase catalyzes the formation of L-DOPA, the rate-limiting step in the biosynthesis of dopamine [22], expression of the tyrosine hydroxylase gene (*th*) becomes a crucial indication of the number of DA neurons. Expression of *th* gene was examined to verify the effect of 6-OHDA on the population of DA neurons in the zebrafish brain (Figure

2C). Our qPCR data showed that expression of the *th* gene was downregulated by 1–2-fold in the disease model. Therefore, the effect of 6-OHDA on expression of the *th* gene was in good agreement with our microscopy observation that the population of DA neurons decreased significantly.

In mammals, DA neuron injury is a well-known cause of movement disorder. In the present study, treatment with 6-OHDA altered the swimming behavior of zebrafish. Zebrafish larvae were exposed to 6-OHDA from 3 dpf to 7 dpf. Typical swimming paths of the control and 6-OHDA-treated zebrafish larvae obtained by a video recording system called Zebrabox are shown in Figure 3A.

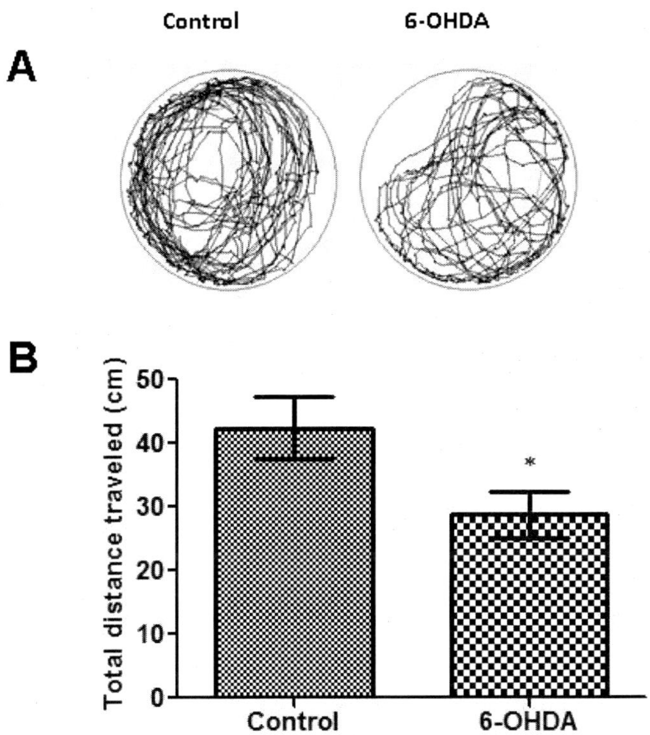

Figure 3. The swimming behavior of zebrafish at 7 dpf following exposure to 6-OHDA for 4 days. (A) Typical swimming patterns of the control and 6-OHDA-treated zebrafish. The lines show the pathway of movement. Zebrafish treated with 6-OHDA were less active than control fish. (B) Quantitative analysis of total distance traveled; * $p < 0.05$.

The swimming paths of the control fish were composed of more complicated lines than that of the 6-OHDA-treated zebrafish, indicating a greater degree of mobility. The total distance traveled decreased significantly ($p < 0.05$) from 42.36 ± 4.9 cm to 28.67 ± 3.7 cm after exposure to 6-OHDA (Figure 3B). Thus, zebrafish larvae exposed to 6-OHDA displayed an adverse impact on swimming behavior after completion of neurodevelopment (7 dpf).

Quantification of Protein Expression in Zebrafish Embryos

The iTRAQ 2D LC-MS/MS method has the advantage that the quantitative relationship of protein regulation between normal and perturbed conditions can be reflected in terms of an average ratio calculated from the ratios of the peptides that contribute to the identification of the protein. On the basis of the Progroup Algorithm function of Protein Pilot, only proteins with an unused Prot score >1.3 (with a confidence over 95%) were selected as detected entries. The relative quantification of a protein was achieved by comparing the peak area of the reported ions derived from the isobaric tag of iTRAQ with molecular mass of 113, 114, 118 and 121. Figure 4 shows a typical tandem MS (MS/MS) fragmentation of peptide GYQYLFECGDYR with a confidence of 99% contributing to the quantification of the protein crystallin β-B1.

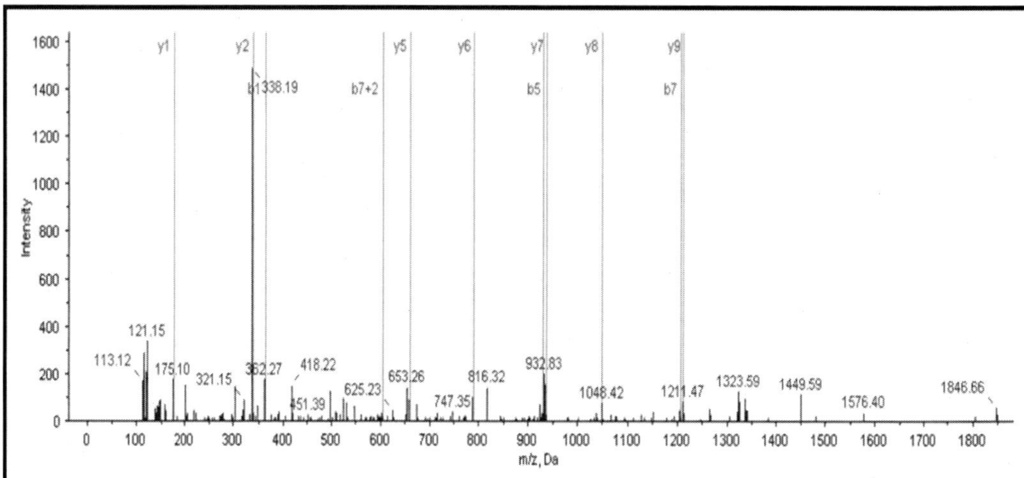

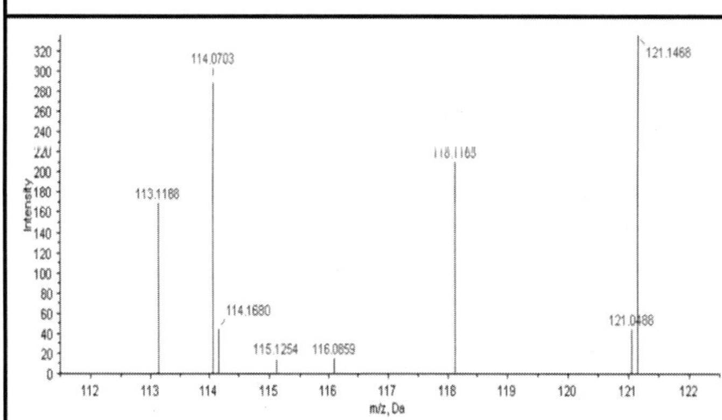

Residue	b	b+2	y	y+2
G	**362.2341**	181.6207	1863.8347	932.4210
Y	525.2974	263.1524	1502.6079	751.8076
Q	653.3560	327.1816	1339.5446	670.2759
Y	816.4193	408.7133	**1211.4860**	606.2466
L	**929.5034**	465.2553	**1048.4227**	524.7150
F	1076.5718	538.7895	**935.3386**	468.1729
E	**1205.6144**	**603.3108**	**788.2702**	394.6387
C[MSH]	1354.6113	677.8093	**659.2276**	330.1174
G	1411.6328	706.3200	510.2307	255.6190
D	1526.6597	763.8335	453.2092	227.1083
Y	1689.7230	845.3652	**338.1823**	169.5948
R	1845.8242	923.4157	**175.1190**	88.0631

Figure 4. Representative MS/MS spectrum of the fragmented peptide GYQYLFECGDYR with 99% confidence and its reporter ion intensity (113 and 118) from the control and the 6-OHDA treated (114 and 121) groups. In all, 24 spectra were used for quantification of the protein.

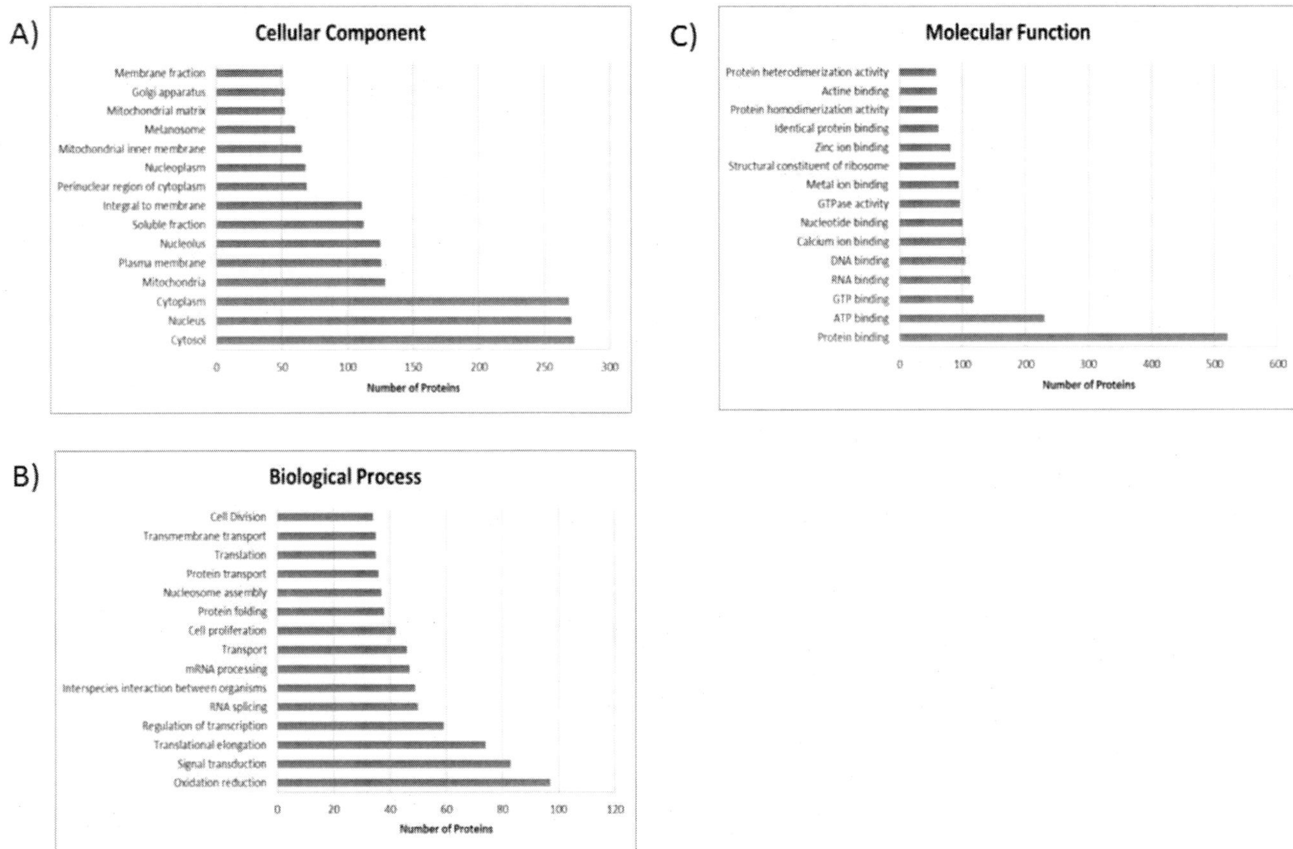

Figure 5. Representative bar graphs of gene ontology (GO) terms for the 1559 unique proteins identified from the combination of control and 6-OHDA samples by LC-MS/MS analysis. (A) Cellular component; (B) biological process; and (C) molecular function.

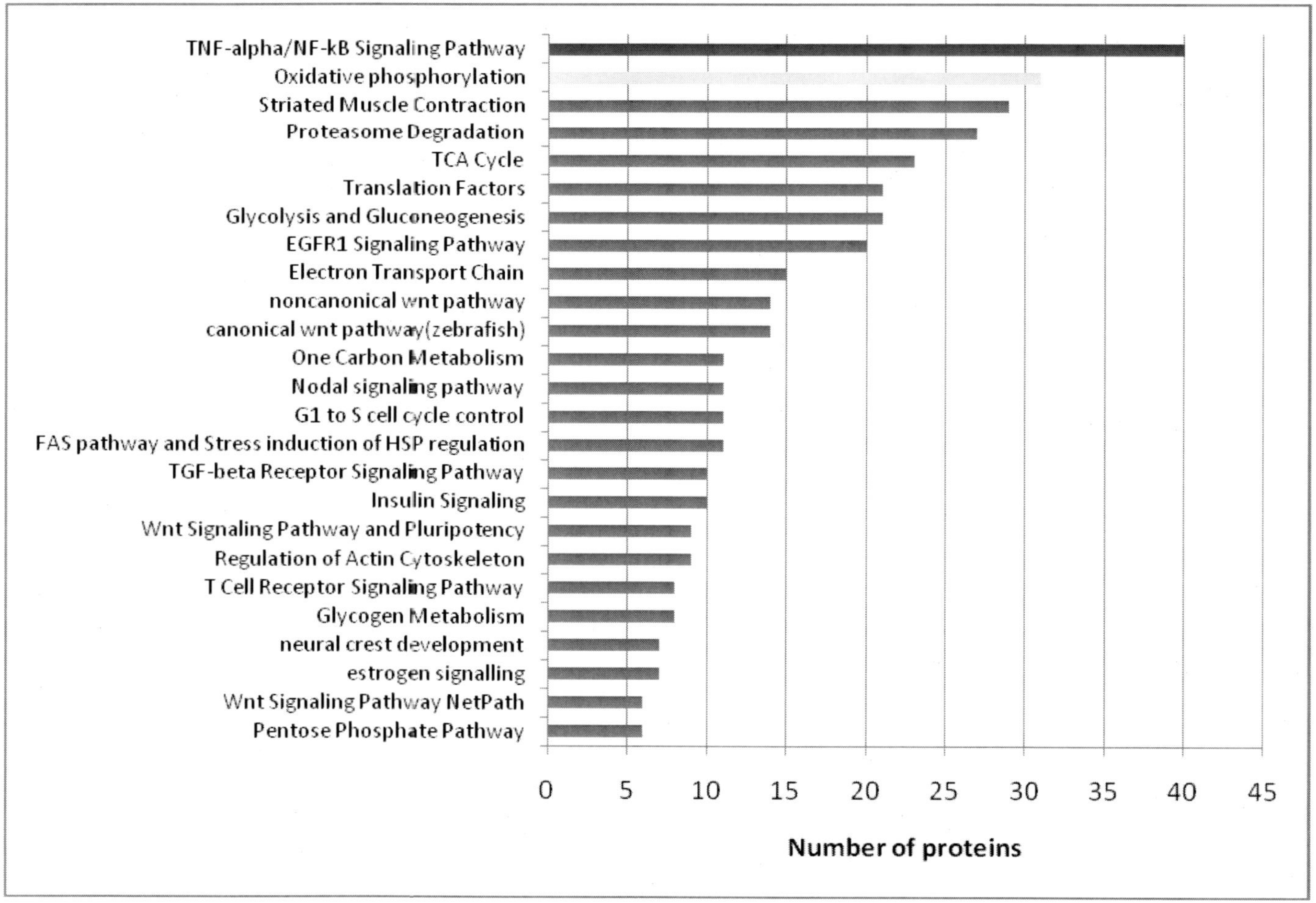

Figure 6. Canonical pathways associated with our proteome data. TNF-α/NF-κB signaling pathway (red) and oxidative phosphorylation of zebrafish (yellow) were connected with the greatest number of detected proteins in the proteome data.

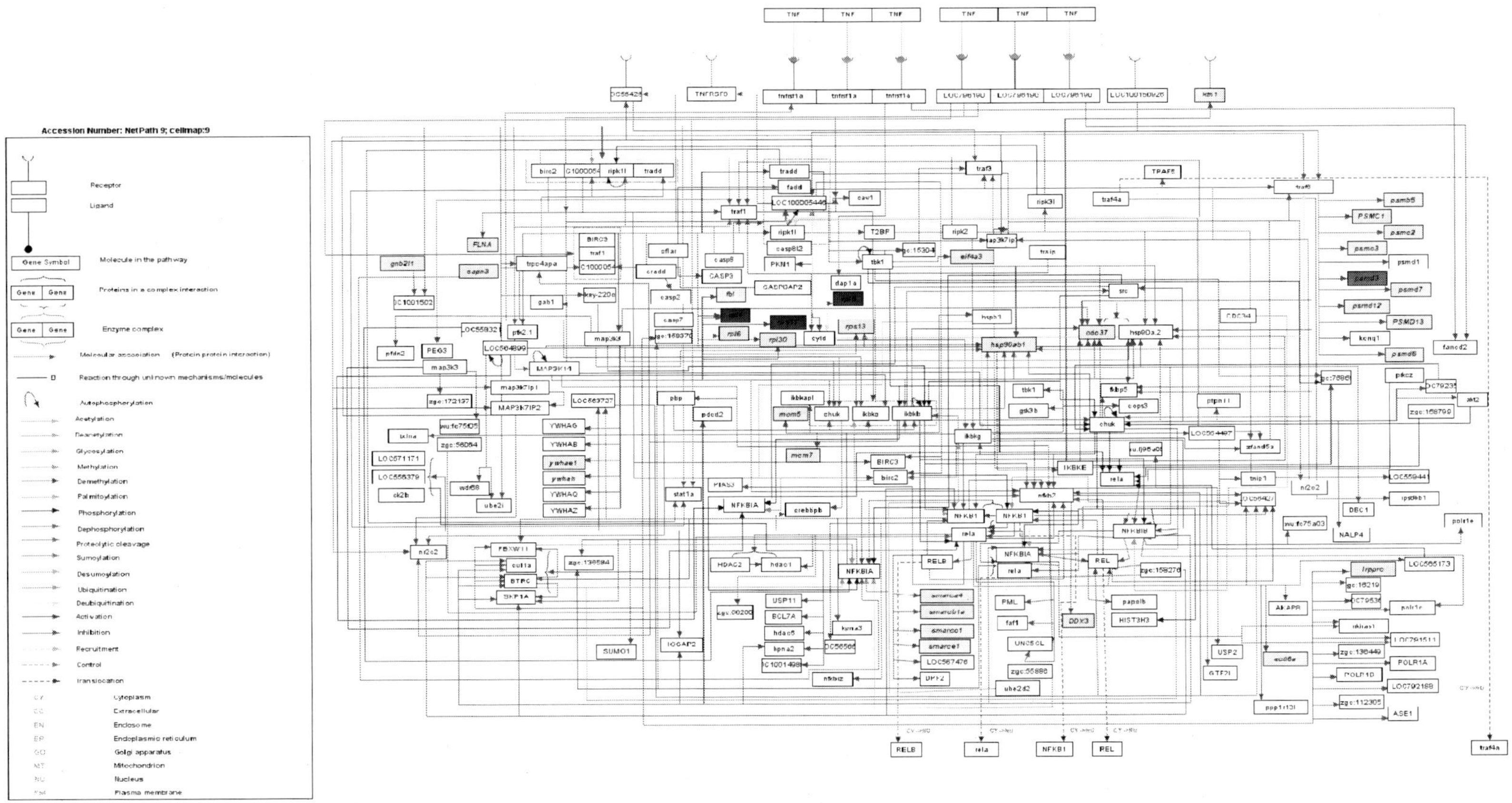

Figure 7. (Continued).

B

Accession No.	Protein name	Fold change (log2)	P value
IPI:IPI00507046.1	psmd3 Proteasome (Prosome, macropain) 26S subunit, non-ATPase, 3	1.08	0.028
IPI:IPI00505526.1	rpl4 Ribosomal protein L4	-0.83	0.000
IPI:IPI00492036.4	rps11 40S ribosomal protein s11	-0.74	0.001
IPI:IPI00505170.3	rpl8 60S ribosomal protein L8	-0.93	0.011

C

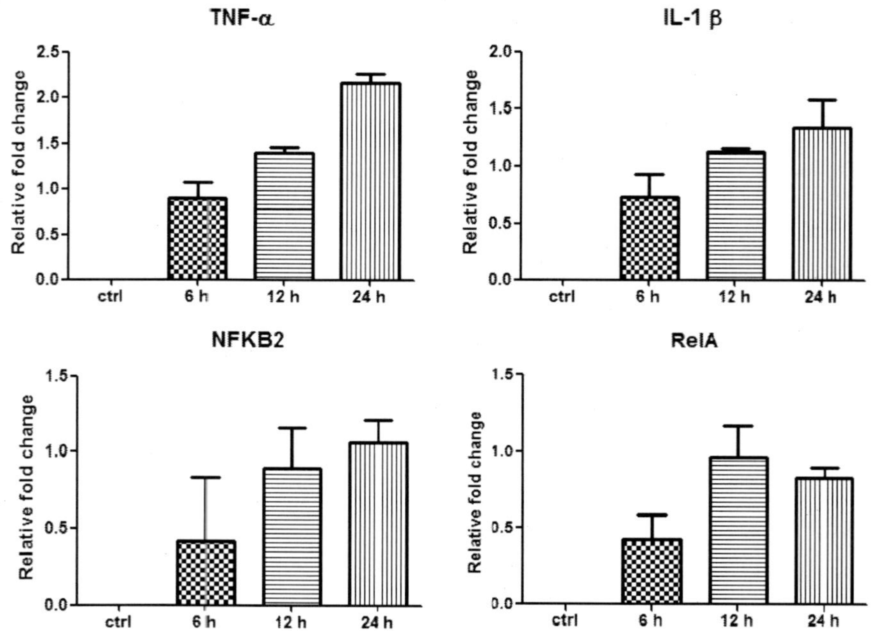

Figure 7. The TNF-α/NF-κB signaling pathway is involved in 6-OHDA-induced neuronal damage in zebrafish. (A, B) Proteins involved in TNF-α/NF-κB signaling pathway identified from the entire proteome data (red, increase of protein abundance; blue, reduction of protein abundance; yellow, no significant change). (C) The relative gene expression level of 6-OHDA-induced pro-inflammatory genes.

Identification of Proteins

A total of 1709 unique proteins (cutoff >1.3) were identified from both control and 6-OHDA treated samples in two independent experiments: 1559 were found to be associated with known gene ontology (GO) terms; 814 (48%) of these annotated proteins were located in the cytosol (273), nucleus (272) and cytoplasm (269) (Figure 5A). The biological processes that contained the greatest number of proteins included oxidation/reduction (98), signal transduction (82) and translational elongation (73) (Figure 5B). The proteins identified were involved in other biological processes, such as regulation of transcription (88), RNA splicing (9) and mRNA processing (7). When categorized by molecular functions, the greatest numbers of proteins were characterized as having protein binding (521) and ATP binding properties (230) (Figure 5C).

Thus, we identified a diverse range of proteins from the control and 6-OHDA-treated zebrafish samples.

Involvement of the TNF-α/NF-κB Signaling Pathway

To access the effect of 6-OHDA on protein expression, we matched all differentially expressed proteins between control and 6-OHDA-treated zebrafish with the corresponding pathways using pathway information from Wikipathways [23] (Figure 6). TNF-α/NF-κB signaling pathway and oxidative phosphorylation pathway were the top candidates among all pathways.

As shown in Figure 7A, the TNF-α/NF-κB signaling pathway was linked with 34 proteins of our proteome (Supplementary material Table 2), with only 4 genes significantly regulated by 6-OHDA (Figure 7B).

After the Wikipathways analysis, only one protein (psmd3, 1.08 fold) was identified with a significant increase in protein abundance (Figure 7). Some other proteins, including rpl4, rps11 and *rpl8* (–0.83, –0.74 and –0.93 fold respectively), exhibited a significant decrease, whereas other proteins, such as *FLNA* and *capn3*, displayed no significant change of protein abundance.

Both TNF-α and NF-κB have been demonstrated to be involved in neuroinflammation. Although neuroinflammation might have initial protective effects, chronic neuroinflammation might promote prolonged neurodegeneration, such as that seen in PD, possibly *via* increased oxidative stress [24]. To validate our finding, some pro-inflammatory genes involved in the TNF-α/NF-κB signaling pathway were further assessed by real-time PCR. As shown in Figure 7C, 6-OHDA stimulated expression of the TNF-α and IL-β genes as well as the NF-κB subunits NFKB2 and RelA, which are key targets of the TNF-α/NF-κB signaling pathway, in a time-dependent manner; however, these targets were not identified in our proteome data. Although SCX-RP 2D LC-MS/MS is regarded as state of the art technology for shotgun proteomics, it gives only limited coverage of global protein expression profile, equivalent to less than <10% of annotated genes in the zebrafish genome.

A

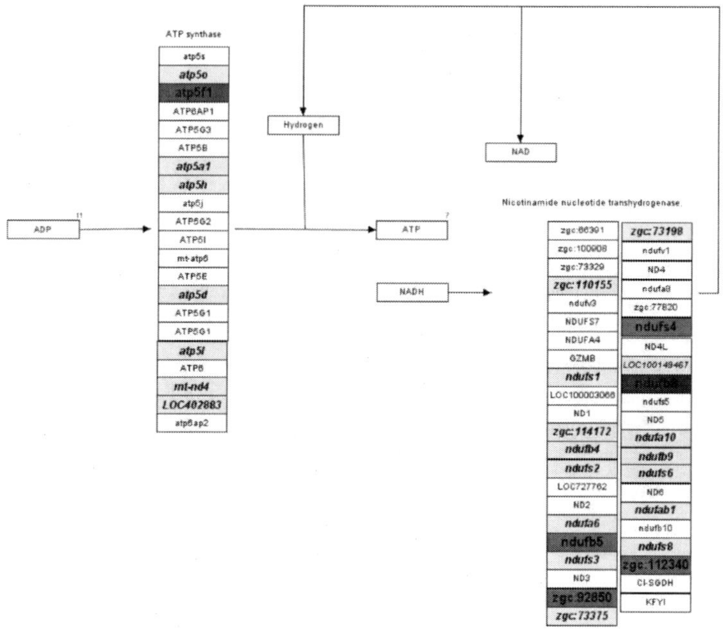

B

Accession No.	Protein name	Fold change (log2)	P value
IPI:IPI00497505.2	ndufs4 Novel protein	0.23	0.011
IPI:IPI00832784.1	ndufb5 NADH dehydrogenase (Ubiquinone) 1 beta subcomplex 5	0.24	0.030
IPI:IPI00513660.1	Zgc:92850	0.25	0.005
IPI:IPI00493072.2	Zgc:112340 Zgc:112340	1.11	0.047
IPI:IPI00496195.2	atp5f1 ATP synthase, H+ transporting, mitochondrial F0 complex, subunit b, isoform 1	-0.42	0.033
IPI:IPI00865737.1	ndufb8 21 kDa protein	-0.92	0.006

Figure 8. (A) Change in protein abundance in oxidative phosphorylation pathway following treatment with 6-OHDA. (B) A list of regulated proteins in oxidative phosphorylation pathway (red, increased protein abundance; blue, reduced protein abundance; yellow, no significant change), $p < 0.05$.

Involvement of the Oxidative Phosphorylation Pathway

The oxidative phosphorylation pathway of zebrafish was connected to 28 proteins (Supplementary material Table 2) and was the top candidate among all associated pathways with differentially expressed proteins. After the Wikipathways analysis, protein *zgc:112340* were identified with a significantly increased protein abundance (Figure 8). Some other proteins, including *atp5f1* (–0.75 fold) and *ndufb8* (–0.53 fold), exhibited a significant decrease, whereas other proteins, such as *atp5h*, *nudfs1* and *zgc:73375*, displayed no significant change of protein abundance. Our results suggest that 6-OHDA modifies the oxidative phosphorylation pathway by regulating the expression level of the above-mentioned target proteins in zebrafish.

DISCUSSION

6-OHDA exerts direct toxicity on neurons and activates inflammatory responses of glial cells. The results of numerous studies support the idea that the glial activation in experimental models of PD, particularly microglia, exacerbates the degeneration of DA neurons [25]. In addition, a low level of dopamine, for instance under PD conditions, requires greater effort for any given movement [26]. Hence, the reduced DA neurons together with altered swimming behavior in the present study demonstrate that 6-OHDA-induced DA neurotoxicity successfully phenocopies PD in zebrafish. Moreover, the qPCR data of *th* gene was in agreement with microscopy observation that 6-OHDA exposure reduced DA neurons, and it also illustrated that the use of whole embryo can pick up signal of relatively DA neuron specific expressed markers, such as tyrosine hydroxylase.

DA neuron loss is not the only prominent symptom of PD, but the subsequent systematic responses, for instance, inflammation and oxidative stress, are also as important. As mentioned above, neuroninflammation might lead to neurodegeneration *via* increased oxidative stress [29]. The glial cells mainly control the inflammatory response in central neural system. Therefore, in several pilot experiments, we tried to apply fluorescence-activated cell (FAC) sorting to isolate GFP^+ glial cells from digested single cell suspension of Tg(gfap:GFP) transgenic zebrafish embryos, in which glial cells are specifically labelled by enhanced green fluorescent protein (EGFP) under the control of the glial fibrillary acidic protein (gfap) promoter [27]. Then FAC-sorted GFP^+ cells were subjected for comparative proteomic profiling to elucidate the mechanism underlying 6-OHDA-induced neurodegeneration. However, finally it failed because of the insufficient amount of proteins from FAC-sorted cells unmet the iTRAQ-2D LC-MS/MS analysis we applied. Several successful studies on gene-profiling of FAC-sorted cells from zebrafish have been reported [28-30]; however, to our knowledge, there has been no report on protein-profiling of FAC-sorted cells from zebrafish so far. For the best of our knowledge, FAC-sorting technique and sensitivity limitation of iTRAQ-2D LC-MS/MS analysis have to be improved in future studies.

Moreover, to have a more complete picture of neurodegenerative diseases, systematic response should also be considered, and there is a need to carry out whole embryo proteome study instead of organ-specific study, such as brain proteome study. Furthermore, we can try to identify and characterize some neural-specific marker with significant changes from whole tissue proteomes. Thus, in the present study it is worth carrying out whole embryo proteome profiling.

Out of 1,709 identified unique proteins from the entire proteome (control and 6-OHDA groups), there were 129 mitochondrial proteins detected. Comparing to other whole embryo proteomic studies, we have identified a relative huge amount of mitochondrial proteins (8% vs 2%) [14]. Such a difference might be due to the exposure of 6-OHDA. Some studies have indicated that mitochondrial dysfunction plays a major role in the onset of PD [31]. Under the neurodegenerative effect of 6-OHDA, the function of mitochondria could also be impaired [32]. It is possible that the dysfunction of mitochondria caused by 6-OHDA affected the number of mitochondrial proteins, although the mechanism behind may require further investigation. Therefore, the identification of huge amount of mitochondrial proteins implied that 6-OHDA might well imitate the disease conditions of PD in terms of mitochondria dysfunction.

Besides, the GO distribution of "Biological Process" category raised a thought that oxidative phosphorylation could be the major reason to the large number of identified oxidation/reduction proteins. Oxidative phosphorylation is a metabolic pathway that uses energy released by oxidation of nutrients to produce adenosine triphosphate (ATP), and it involves a series of redox reactions that are carried out by the protein complex within mitochondria. In fact, PD is often linked to the dysregulation of oxidative phosphorylation [31-34]. So, it is possible that the neurodegenerative effect of 6-OHDA was reflected on the number of oxidation/reduction proteins due to dysfunction of mitochondria. Furthermore, the assumption was reassured by the huge portion of ATP binding proteins under "Molecular functions" category as oxidative phosphorylation is highly associated with ATP production. In summary, the analysis of GO term distribution implied the neurodegenerative effect of 6-OHDA on mitochondria and oxidative phosphorylation pathway.

Tumor necrosis factor-α (TNF-α) and nuclear factor κ–light chain enhancer of activated B cells (NF-κB) are highly associated with our entire proteome. Our qPCR data of some major targets, like TNF-α and IL-β as well as the NF-κB subunits NFKB2 and RelA, demonstrated the involvement of TNF-α/NF-κB pathway under 6-OHDA-induced neurodegenerative effect, though these targets were not identified in our proteome. The inability to identify key targets also illustrated the limitation of shotgun proteomics employed in the present study. The limited coverage only provides less than 10% of zebrafish genome annotation database. Nevertheless, the involvement of TNF-α/NF-κB pathway is not surprising because they both have long been associated with neuroinflammation and PD.

On one hand, TNF-α, a multifunctional proinflammatory cytokine, has been demonstrated to play a major role in the CNS neuroinflammation-mediated cell death in Alzheimer's disease (AD), PD and amyotrophic lateral sclerosis (ALS) as well as several other CNS complications [35]. Studies indicated that chronic expression of high levels of TNF-α induces neuronal loss and a progressive neurodegenerative effect in the *substantial nigra* of adult mice [36, 37]. On the other hand, NF-κB is a protein complex that controls the transcription of DNA. Similar to TNF-α, the incorrect regulation of NF-κB has been linked to inflammatory diseases [38-40]. Emerging evidence suggests that NF-κB is activated within the CNS under many pathological settings, including neurodegenerative diseases [33, 34]. Though we might have failed to identify some key target genes, the close affinity between TNF-α/NF-κB signaling pathway and our proteome validated the neurodegenerative effect of 6-OHDA.

On the basis of the Wikipathways analysis, six genes of the oxidative phosphorylation pathway were being regulated following exposure to 6-OHDA (Figure 8). Among all these differentially expressed proteins, NADH dehydrogenase is the protein affected most in the oxidative phosphorylation pathway (5 out of 6 genes). NADH dehydrogenase is located in the inner mitochondrial membrane that catalyzes the transfer of electrons from NADH to coenzyme Q. NADH dehydrogenase also produces superoxide, a reactive oxygen species that contributes to oxidative stress. Superoxide is not particularly reactive but it can initiate lipid peroxidation in the hydroperoxyl form and result in cell damage. It has been hypothesized that the overproduction of ROS caused by NADH dehydrogenase defects is involved in the animal model of PD [41, 42]. Some studies have shown that inhibition of NADH dehydrogenase by rotenone stimulates the production of ROS [43,44]; under PD conditions, therefore, the NADH dehydrogenase defect could play a key role in mitochondria dysfunction, or the consequential cell death, due to the overproduction of ROS. Indeed, accumulation of oxidative damage can affect the efficiency of mitochondria and further increase the rate of ROS production [45].

More importantly, identification of the oxidative phosphorylation pathway matched with our Blast2GO analysis, where oxidation/reduction and ATP binding occupied a significant portion of "Biological Process" and "Molecular Function" (Figure 5B and C). Moreover, 6-OHDA appears to be toxic to the mitochondrial complex I [46, 47]. Because oxidative phosphorylation is responsible for producing ATP, it is possible that proteins related to ATP are regulated under the effect of 6-OHDA.

Combining all the available evidence, we demonstrated the neurotoxicity of 6-OHDA in zebrafish but our data show that there is a possibility that the neurotoxin modifies the function of NADH dehydrogenase and leads to an overproduction of ROS. Therefore, the dysfunction of NADH dehydrogenase might play an important role in triggering the 6-OHDA-induced neurodegenerative effect in the zebrafish model.

We have presented a comprehensive study using a combination of *in vivo* imaging, real-time qPCR and MS-based proteomics to demonstrate the 6-OHDA-induced neurodegeneration in the zebrafish model. The emphasis of the present study is on monitoring the quantitative global expression change in order to provide evidence for the molecular mechanism underlying the neuronal damage. This phenomenon is unclear in much of the literature and it will need to be firmly established before further pharmacological studies can be undertaken.

Neuronal damage could originate from different cellular pathways in the complex regulatory network and an *in vivo* study could provide considerably more information than *in vitro* assays using purified targets. Proteins are the primary molecules that carry out biological functions in cells and the ability to examine the proteome of an organism provides a robust overview of the physiological changes that occur and could greatly augment target-oriented biological data. By comparison, the methodology of immunology depends on the availability of antibodies and *a priori* knowledge of the targets and is generally not amenable to global monitoring.

Proteomics has advantages over transcriptomics, such as the microarray, in terms of observation of functional change. Microarrays, although large-scale and high-throughput, only infer functional changes from RNA transcript concentration. It is now known that transcript level often does not correlate with protein expression level; microarrays thus provide less direct information about functional expression changes. Earlier studies mentioned that microarray results show much more pronounced effects on developmental processes compared to the proteomic results. However, the proteomic approach defines responses at the protein level that were probably not regulated at the transcription level, thus providing additional information [48].

We evaluated the results obtained by the method described here by comparison with a protein database of zebrafish. For example, by comparing Gene Ontology (GO) terms under cellular components, we found that membrane proteins were more appropriately represented in our proteome. As Figure 4A shows, hydrophobic species, such as membrane proteins (e.g. plasma membrane, integral to membrane and mitochondrial inner membrane) are a significant portion of cellular components. The GO term membrane comprised 13% (547 out of 3,928) of the entire cellular components category (Figure 5A). Upon comparing the zebrafish proteome from the present study against the brain [49] and whole embryo [14] datasets obtained from 2-D gel and MALDI-TOF MS/MS analysis, it was found that ~3% and ~0.6% of membrane proteins were present among them, respectively. Membrane proteins are commonly under-represented in 2-D gel proteomics analysis. One of the ways to increase the proteome depth would be to modify sample preparation to selectively enrich membrane proteins.

Nevertheless, the bigger issue is that the under-representation of certain types of proteins might eventually lead to the over-representation of the others. Indeed, cytoplasm proteins are usually over-represented in the 2-D gel method. The 2-D gel method can sometimes contain >80% of cytoplasm proteins as compared to the 30% in our data (1,171 out of 3,928)[14, 49]. Therefore, the iTRAQ-2D LC-MS/MS technique is able to provide a wider range and more reasonable set of proteome data. The proteome obtained from this study can be compared for understanding the proteomic changes due to neuronal damage and PD.

Another important finding of this study was that the proteome profile of 6-OHDA indeed reflects biological processes that are related to PD. On one hand, TNF-α and NF-κB are proven to have a close connection with cell inflammation and neuronal death, though the exact mechanisms remain unclear and controversial, particularly in the zebrafish model. On the other hand, oxidative phosphorylation dysregulation and an NADH dehydrogenase defect could lead to overproduction of ROS, which is suspected to be one of the major causes of PD. To the best of our knowledge, 6-OHDA is one of the few neurotoxins that imitate PD conditions by selectively killing DA neurons. It will be interesting to understand the details of how the biological pathways identified in this study and discussed above interact and mediate the neurodegenerative effects of 6-OHDA in zebrafish.

ACKNOWLEDGMENTS

This work was supported by a grant from the Science and Technology Development Fund of Macau SAR (reference number 045/2007/A3 and 078/2011/A3) and the Research Committee, University of Macau (reference numberUL017/09-Y1).

REFERENCES

[1] Beal MF: Experimental models of Parkinson's disease. *Nature reviews* 2001, 2:325-334.

[2] Blum D, Torch S, Lambeng N, Nissou M, Benabid AL, Sadoul R, Verna JM: Molecular pathways involved in the neurotoxicity of 6-OHDA, dopamine and MPTP: contribution to the apoptotic theory in Parkinson's disease. *Progress in neurobiology* 2001, 65:135-172.

[3] Luthman J, Fredriksson A, Sundstrom E, Jonsson G, Archer T: Selective lesion of central dopamine or noradrenaline neuron systems in the neonatal rat: motor behavior and monoamine alterations at adult stage. *Behavioural brain research* 1989, 33:267-277.

[4] Sachs C, Jonsson G: Mechanisms of action of 6-hydroxydopamine. *Biochemical pharmacology* 1975, 24:1-8.

[5] Perumal AS, Gopal VB, Tordzro WK, Cooper TB, Cadet JL: Vitamin E attenuates the toxic effects of 6-hydroxydopamine on free radical scavenging systems in rat brain. *Brain research bulletin* 1992, 29:699-701.

[6] Kumar R, Seth RK, Sekhon MS, Bhargava JS: Serum lipid peroxide and other enzyme levels of patients suffering from thermal injury. *Burns* 1995, 21:96-97.

[7] Sherer TB, Betarbet R, Stout AK, Lund S, Baptista M, Panov AV, Cookson MR, Greenamyre JT: An in vitro model of Parkinson's disease: linking mitochondrial impairment to altered alpha-synuclein metabolism and oxidative damage. *J. Neurosci.* 2002, 22:7006-7015.

[8] Hasegawa E, Takeshige K, Oishi T, Murai Y, Minakami S: 1-Methyl-4-phenylpyridinium (MPP+) induces NADH-dependent superoxide formation and enhances NADH-dependent lipid peroxidation in bovine heart submitochondrial particles. *Biochemical and biophysical research communications* 1990, 170:1049-1055.

[9] Hanash S: Disease proteomics. *Nature* 2003, 422:226-232.

[10] Datta A, Park JE, Li X, Zhang H, Ho ZS, Heese K, Lim SK, Tam JP, Sze SK: Phenotyping of an in vitro model of ischemic penumbra by iTRAQ-based shotgun quantitative proteomics. *Journal of proteome research* 2010, 9:472-484.

[11] Li ZH, Alexa D, Siu SO, Chu IK, Renn J, Winkler C, Zhao HY, Yan WR, Mahady GB, Li GH, et al: Combined in vivo imaging and omic approaches reveal metabolism of icaritin and its glycosides in zebrafish larvae. *Molecular BioSystems* 2011, accepted

[12] Gygi SP, Rist B, Gerber SA, Turecek F, Gelb MH, Aebersold R: Quantitative analysis of complex protein mixtures using isotope-coded affinity tags. *Nature biotechnology* 1999, 17:994-999.

[13] Lucitt MB, Price TS, Pizarro A, Wu W, Yocum AK, Seiler C, Pack MA, Blair IA, Fitzgerald GA, Grosser T: Analysis of the zebrafish proteome during embryonic development. *Mol. Cell Proteomics* 2008, 7:981-994.

[14] Tay TL, Lin Q, Seow TK, Tan KH, Hew CL, Gong Z: Proteomic analysis of protein profiles during early development of the zebrafish, Danio rerio. *Proteomics* 2006, 6:3176-3188.

[15] Lewis KE, Eisen JS: From cells to circuits: development of the zebrafish spinal cord. *Progress in neurobiology* 2003, 69:419-449.

[16] Norrby K: In vivo models of angiogenesis. *Journal of cellular and molecular medicine* 2006, 10:588-612.

[17] Zhang ZJ, Cheang LC, Wang MW, Lee SM: Quercetin exerts a neuroprotective effect through inhibition of the iNOS/NO system and pro-inflammation gene expression in PC12 cells and in zebrafish. *International journal of molecular medicine* 2011, 27:195-203.

[18] Livak KJ, Schmittgen TD: Analysis of relative gene expression data using real-time quantitative PCR and the 2(-Delta Delta C(T)) Method. *Methods (San Diego, Calif* 2001, 25:402-408.

[19] Link V, Shevchenko A, Heisenberg CP: Proteomics of early zebrafish embryos. *BMC developmental biology* 2006, 6:1.

[20] Rink E, Wullimann MF: The teleostean (zebrafish) dopaminergic system ascending to the subpallium (striatum) is located in the basal diencephalon (posterior tuberculum). *Brain Res* 2001, 889:316-330.

[21] Rink E, Wullimann MF: Connections of the ventral telencephalon and tyrosine hydroxylase distribution in the zebrafish brain (Danio rerio) lead to identification of an ascending dopaminergic system in a teleost. *Brain research bulletin* 2002, 57:385-387.

[22] Haavik J, Toska K: Tyrosine hydroxylase and Parkinson's disease. *Molecular neurobiology* 1998, 16:285-309.

[23] Pico AR, Kelder T, van Iersel MP, Hanspers K, Conklin BR, Evelo C: WikiPathways: pathway editing for the people. *PLoS Biol.* 2008, 6:e184.

[24] Whitton PS: Inflammation as a causative factor in the aetiology of Parkinson's disease. *British journal of pharmacology* 2007, 150:963-976.

[25] Przedborski S: Clarification: Pathogenic role of glial cells in Parkinson's disease. *Mov. Disord.* 2004, 19:118.

[26] Obeso JA, Rodriguez-Oroz MC, Benitez-Temino B, Blesa FJ, Guridi J, Marin C, Rodriguez M: Functional organization of the basal ganglia: therapeutic implications for Parkinson's disease. *Mov. Disord.* 2008, 23 Suppl 3:S548-559.

[27] Kassen SC, Ramanan V, Montgomery JE, C TB, Liu CG, Vihtelic TS, Hyde DR: Time course analysis of gene expression during light-induced photoreceptor cell death and regeneration in albino zebrafish. *Developmental neurobiology* 2007, 67:1009-1031.

[28] Covassin L, Amigo JD, Suzuki K, Teplyuk V, Straubhaar J, Lawson ND: Global analysis of hematopoietic and vascular endothelial gene expression by tissue specific microarray profiling in zebrafish. *Developmental biology* 2006, 299:551-562.

[29] Cerda GA, Hargrave M, Lewis KE: RNA profiling of FAC-sorted neurons from the developing zebrafish spinal cord. *Dev. Dyn.* 2009, 238:150-161.

[30] Stuckenholz C, Lu L, Thakur P, Kaminski N, Bahary N: FACS-assisted microarray profiling implicates novel genes and pathways in zebrafish gastrointestinal tract development. *Gastroenterology* 2009, 137:1321-1332.

[31] Jenner P, Olanow CW: The pathogenesis of cell death in Parkinson's disease. *Neurology* 2006, 66:S24-36.

[32] Tirmenstein MA, Hu CX, Scicchitano MS, Narayanan PK, McFarland DC, Thomas HC, Schwartz LW: Effects of 6-hydroxydopamine on mitochondrial function and glutathione status in SH-SY5Y human neuroblastoma cells. *Toxicol. In Vitro* 2005, 19:471-479.

[33] Nakai M, Qin ZH, Chen JF, Wang Y, Chase TN: Kainic acid-induced apoptosis in rat striatum is associated with nuclear factor-kappaB activation. *Journal of neurochemistry* 2000, 74:647-658.

[34] Mattson MP, Camandola S: NF-kappaB in neuronal plasticity and neurodegenerative disorders. *The Journal of clinical investigation* 2001, 107:247-254.

[35] Greig NH, Mattson MP, Perry T, Chan SL, Giordano T, Sambamurti K, Rogers JT, Ovadia H, Lahiri DK: New therapeutic strategies and drug candidates for neurodegenerative diseases: p53 and TNF-alpha inhibitors, and GLP-1 receptor agonists. *Annals of the New York Academy of Sciences* 2004, 1035:290-315.

[36] Swarup V, Das S, Ghosh S, Basu A: Tumor necrosis factor receptor-1-induced neuronal death by TRADD contributes to the pathogenesis of Japanese encephalitis. *Journal of neurochemistry* 2007, 103:771-783.

[37] Chertoff M, Di Paolo N, Schoeneberg A, Depino A, Ferrari C, Wurst W, Pfizenmaier K, Eisel U, Pitossi F: Neuroprotective and neurodegenerative effects of the chronic expression of tumor necrosis factor alpha in the nigrostriatal dopaminergic circuit of adult mice. *Experimental neurology* 2011, 227:237-251.

[38] Albensi BC, Mattson MP: Evidence for the involvement of TNF and NF-kappaB in hippocampal synaptic plasticity. *Synapse (New York, NY* 2000, 35:151-159.

[39] Ha SK, Moon E, Kim SY: Chrysin suppresses LPS-stimulated proinflammatory responses by blocking NF-kappaB and JNK activations in microglia cells. *Neuroscience letters* 2010, 485:143-147.

[40] Singh K, Chaturvedi R, Asim M, Barry DP, Lewis ND, Vitek MP, Wilson KT: The apolipoprotein E-mimetic peptide COG112 inhibits the inflammatory response to Citrobacter rodentium in colonic epithelial cells by preventing NF-kappaB activation. *The Journal of biological chemistry* 2008, 283:16752-16761.

[41] Petroske E, Meredith GE, Callen S, Totterdell S, Lau YS: Mouse model of Parkinsonism: a comparison between subacute MPTP and chronic MPTP/probenecid treatment. *Neuroscience* 2001, 106:589-601.

[42] Schapira AH, Cooper JM, Dexter D, Jenner P, Clark JB, Marsden CD: Mitochondrial complex I deficiency in Parkinson's disease. *Lancet* 1989, 1:1269.

[43] Hensley K, Pye QN, Maidt ML, Stewart CA, Robinson KA, Jaffrey F, Floyd RA: Interaction of alpha-phenyl-N-tert-butyl nitrone and alternative electron acceptors with complex I indicates a substrate reduction site upstream from the rotenone binding site. *Journal of neurochemistry* 1998, 71:2549-2557.

[44] Seaton TA, Cooper JM, Schapira AH: Free radical scavengers protect dopaminergic cell lines from apoptosis induced by complex I inhibitors. *Brain Res.* 1997, 777:110-118.

[45] Betarbet R, Sherer TB, MacKenzie G, Garcia-Osuna M, Panov AV, Greenamyre JT: Chronic systemic pesticide exposure reproduces features of Parkinson's disease. *Nature neuroscience* 2000, 3:1301-1306.

[46] Betarbet R, Sherer TB, Di Monte DA, Greenamyre JT: Mechanistic approaches to Parkinson's disease pathogenesis. *Brain pathology (Zurich, Switzerland)* 2002, 12:499-510.

[47] Cleeter MW, Cooper JM, Schapira AH: Irreversible inhibition of mitochondrial complex I by 1-methyl-4-phenylpyridinium: evidence for free radical involvement. *Journal of neurochemistry* 1992, 58:786-789.

[48] De Wit M, Keil D, van der Ven K, Vandamme S, Witters E, De Coen W: An integrated transcriptomic and proteomic approach characterizing estrogenic and metabolic effects of 17 alpha-ethinylestradiol in zebrafish (Danio rerio). *General and comparative endocrinology* 2010, 167:190-201.

[49] Singh SK, Sundaram CS, Shanbhag S, Idris MM: Proteomic profile of zebrafish brain based on two-dimensional gel electrophoresis matrix-assisted laser desorption/ionization MS/MS analysis. *Zebrafish* 2010, 7:169-177.

In: Biochemistry and Molecular Biology Coming of Age
Editor: Mary Miu Yee Waye
ISBN: 978-1-63321-815-4

Chapter 7

USING SYNTHETIC LETHALITY IN THE TARGETED TREATMENT OF HUMAN CANCERS

Matthew J. Imrie**, *Alexandria Jack and Francis Choy
Department of Biology, Centre for Biomedical Research,
University of Victoria, Station CSC, Victoria, BC, Canada

ABSTRACT

Chemotherapeutics are globally accepted drugs for treating cancerous malignancies. The discovery of new anti-cancer drugs is generally not dependent on the genetic context of gene interactions within the cell. Additionally, chemotherapeutics have negative effects on non-cancerous cells. Gene dosage can be used as a basis to explain the concepts of oncogene addiction (the apparent dependency of cancers on one or a few genes for the expression and maintenance of the malignant phenotype) and synthetic lethality. Two genes are said to be synthetically lethal if a deleterious mutation in both genes results in cell death. They are said to be synthetically sick if their deletion results in severely reduced fitness. We discuss the synthetic lethality surrounding the DNA repair protein poly(ADP-ribose) polymerase and its interactions with mutations in *BRCA1*, *BRCA2* and *mTOR*. We then explain how one of the hallmarks of cancer, an abundance of mutations, was used to determine potential synthetic lethal interactions in human cancers using investigations into interactions between homologous gene products from *Saccharomyces cerevisiae*. Large scale models such as the ones described here could have a significant impact on future investigations into potential targets for anti-cancer drug therapies.

INTRODUCTION

Cancer is a disease caused by mutations; current data found on the Catalogue of Somatic Mutations in Cancer (COSMIC) (http://cancer.sanger.ac.uk/cancergenome/projects/census/,

* Departments of Computer Science and Biology, University of Victoria, PO Box 3020 Station CSC, Victoria, BC, V8W 3N5 Email: imrie@cs.uvic.ca.

Forbes et al., 2008. Accessed August 2012) put the total number of genes causally related to cancer ("census" genes) at 487. Both somatic and germ line mutations have been found within these census genes, including base substitutions, translocations, and frame shift mutations. Furthermore, 233,349 individual mutations have been identified over 20, 000 genes as occurring in cancerous cells (Futreal et al., 2004). While this astounding number accounts for the near majority of all known genes in the human genome, this is hardly representative of a single particular cancerous cell line, which may only exhibit a fraction of the catalogued mutations. Regardless of the number of mutations occurring within cancerous cells (an estimated 10^4-10^5 acquired mutations) they are in a precarious position as large scale genetic perturbations affect cellular homeostasis (Ashworth et al., 2011). Many of these perturbations are related to gene dosage levels and mutational changes affecting the interacting partners of gene interaction networks (Ashworth et al., 2011; Hanahan and Weinberg, 2000, 2011; Klein, 1981; Lando et al., 2009; Mendes-Pereira et al., 2009; Scholl et al., 2009).

Gene dosage, and specifically the gene balance hypothesis, is the concept that genes encoding proteins in pathways and multimeric complexes are expressed at specific levels unique to their environment and interaction partners (Birchler et al., 2005; Birchler and Veitia, 2007, 2010). In other words, gene expression is balanced across the genome between all interacting partners, for any given environment or cellular state. Consider an example where regulatory heterodimers form to affect gene expression. An overabundance of one subunit may produce a homodimer, resulting in abnormal expression of affected genes. Such a case are the *Saccharomyces cerevisiae* transcription factors Pip2p and Oaf1p which form a heterodimer responsible for the regulation of expression of peroxisomal proteins (Rottensteiner et al., 1997). *Pip2p*[1] knockouts were shown to form putative Oaf1p homodimers. Such an observation lends credence to the idea that an overexpression of Oaf1p, beyond the required dosage requirements, would result in the formation of a greater percentage of Oaf1p homodimers. This preference to homodimerize over forming Pip2p-Oaf1p heterodimers would alter the desired expression of peroxisomal proteins.

Another example of gene balancing involves the regulation of IL-2 expression in human T lymphocytes by two forms of the transcription factor NF-κB, namely, the p50 subunit homodimer and the p50 subunit and p65 subunit heterodimer. Following an antigenic response, the amount of homodimer decreases, whereas the amount of heterodimer increases. Correlated with heterodimer formation, homodimers are sequestered to the nucleus where they are kept from forming complete NF-κB (Kang et al., 1992). If the dosage level of p50 were increased to out-compete the formation of heterodimers, few, if any, heterodimers would be formed. Altering the dosage of p50 to preferentially create homodimers reduces IL-2 levels (Kang et al., 1992), which could affect the immunological response of T-cells.

When considering gene dosage in cancers, it can be the loss of function of the gene product, rather than overexpression of a particular gene in a pathway that causes gene dosage imbalance. One example of this involves the tumour suppressor genes *BCRA1* and *BCRA2* which are members of the DNA homologous recombination repair (HR) pathway. When these genes are mutated in such a way that their wildtype functions are lost, as they are in breast cancer, there is an imbalance in the HR pathway and subsequently, a major mechanism of the DNA repair repertoire is removed. The result of this loss is an increase in the risk of breast cancer.

---

[1] Abbreviations in *italics* indicate gene names, while non-italicized abbreviations represent the gene products.

The loss of gene function is not relegated to a single or pairwise instance; rather, a repertoire of genes can be affected. In line with cancer's mutational basis, cancerous cell-specific pathways leading to the expression of the malignant phenotype can be arrested/reversed due to a condition known as *oncogene addiction* (Weinstein, 2002; Weinstein and Joe, 2008). This is a condition in which the cancerous cells are dependent upon one or a few particular oncogenes that, when functionally blocked, cause inhibited growth or cell death. In this situation, the mutations in those pathways accumulate, allowing the oncogenic pathway to gain new and unique function within itself, but not with its wild-type partners. The reliance on the expression of an oncogene to maintain a core function essential for the cancerous cells' survival has completely eliminated the ability to buffer the negative impact when one or more components of the oncogene are inhibited. This means that the cancerous cell is solely dependent on the oncogenic pathway – explaining why oncogenic addiction has been referred to as the "Achilles heel of cancer" (Weinstein, 2002).

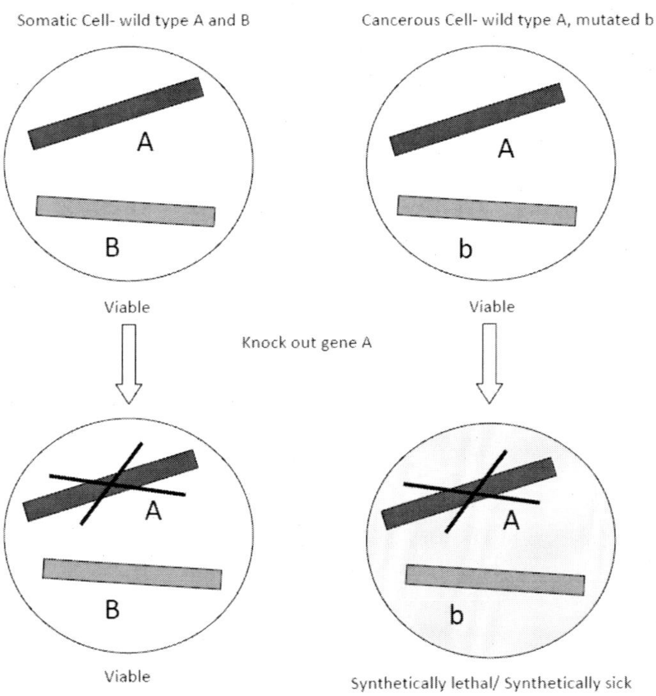

Figure 1. The mechanism of synthetic lethality. Wild type forms of genes are represented by capital letters, whereas lower case represents a mutated form; genes A and B are located on different chromosomes in this example, but could be present on the same chromosome. Gene B is mutated in the cancerous cell but not in the somatic cell, knock out of gene A results in synthetic lethality in the cancerous cells, but does not affect the viability of somatic cells. In the case where gene A is mutated in the cancerous cell line, gene B could be knocked out to create a synthetically lethal phenotype in cancerous cells, as depicted here.

A loss of genetic buffering means that the cell will be unable able to cope with the loss of expression of genes to which they are addicted. Consequentially, this phenomena has created the opportunity to destroy cancerous cells using synthetic lethality (Nijman, 2011) (Figure 1).

Two genes are said to be synthetically lethal if a deleterious mutation in both genes results in cell death or severely reduce fitness (synthetically sick) (Figure 1, the cell in grey, lower right corner). Conversely, if the deletion occurs in one of the two genes, there is only a mild or no reduction in fitness (Figure 1, the cells at the upper right and lower left). As oncogene addiction has effectively disposed of redundancies in the gene interaction network of tumor cells, identifying synthetically lethal genes may provide drug targets that will kill cancer cells, but only mildly perturb healthy cells.

Synthetic lethality could be advantageous to the development of new cancer therapeutics, since a major issue when treating patients with cancer is the toxicity of common therapeutics to specific non-cancerous somatic cells (Earhart et al.1983; Olson and Mushlin 1973; Rudders and Hensley 1973; Stentoft 1990). The utility of these therapeutics is indisputable, as without them patient longevity would be severely reduced. In this chapter, we discuss the following question:

- Could chemotherapeutic targets be found that leverage the mutational effects of cancer, targeting cancerous cells while having little to no effect on healthy cells?

We will first introduce an example of anti-cancer drug discovery based on synthetic lethality, and then discuss attempts to use high throughput gene interaction studies in *S. cerevisiae* to discover novel anti-cancer drug targets. Finally, we outline future directions of work in this field.

Synthetic Lethality in Human Cancer - A Simple Example

Synthetic lethality would appear to be a useful avenue in the search for anti-cancer drug targets; however, no large scale studies aimed at identifying putative synthetically lethal genes have been conducted for genetic interactions within human cells. Synthetically lethal interactions must be found by investigating individual gene interactions. For this reason, only a handful of synthetic genetic interactions are known for human cell lines – approximately 600 (www.thebiogrid.com. Stark et al., 2006. Accessed August 2012). In comparison, it has been estimated that the number of interactions for the human protein-protein interaction network is up to 369,000 (Hart et al., 2006).

As an example, RNA interference (RNAi) screening has been used in the quest to elucidate the synthetically lethal partners of p53 (Berns et al. 2004). In another case, synthetically lethal partners of the most commonly mutated oncogene, the cellular signalling protein KRAS, were queried in cells containing mutant KRAS using RNAi (Scholl et al. 2009). This resulted in the discovery of a synthetically lethal sensitivity to the serine-threonine kinase STK33 (Scholl et al., 2009).

Additionally, a particularly well studied example is the poly(ADP-ribose) polymerase (PARP1) gene and its synthetically lethal partners. In the following sections, we will further discuss PARP1 and its synthetically lethal partners BRCA1, BRCA2, and PTEN (phosphatase and tension homologue protein). When discussing BRCA1 and BRCA2, we will use our understanding of synthetic lethality in PARP1 interactions to introduce the novel PARP1 inhibitor Olaparib.

The Role of PARP1

PARP1 signals DNA repair by binding to breaks in DNA and attracting DNA repair enzymes, initiating base excision repair (Bryant et al. 2005). This repair enzyme has a synthetically lethal relationship with the previously mentioned HR repair enzymes BRCA1 and BRCA2 (Iglehart and Silver, 2009). It also exhibits a synthetically sick relationship with the tumour suppressor PTEN, another one of the most commonly mutated genes in cancer (Li et al. 1997). Additionally, PTEN has been found to have a synthetically sick relationship with the serine/threonine kinase mTOR (mammalian target of rapamycin protein (Neshat et al. 2011). These interactions are depicted in Figure 2.

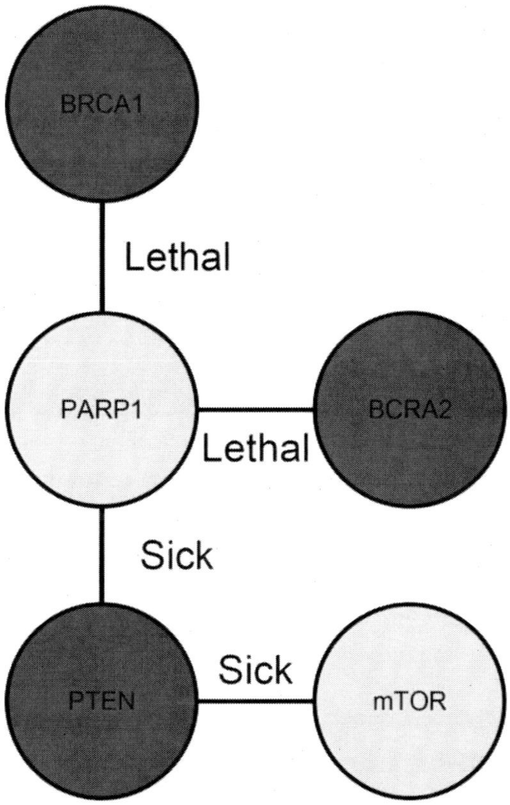

Figure 2. The synthetic lethal interactions of PARP1 and PTEN. Oncogenic mutant genes are colored red. The targets of gene knockdown causing synthetic lethality/sickness are colored blue.

PARP1 and BRCA - A Model for Cancer Drug Design

DNA repair is a major theme in cancer genetics, as DNA repair mechanisms assist in the prevention of mutations, and mutations cause cancer. As previously mentioned, PARP1 is involved in base excision repair; a major pathway in the repair of single stranded DNA breaks.

Additionally, in the absence of BRCA1 and BRCA2, PARP1 acts in a buffering capacity by repairing double stranded breaks (Bryant et al. 2005). In the absence of functional PARP1, spontaneous single stranded breaks in DNA collapse replication forks, creating DNA lesions and initiating the HR pathway. When BRCA1 or BRCA2 function has been lost in a cancerous cell, the HR pathway is unable to function and the DNA lesions persist and accumulate (Bryant et al. 2005). This demonstrates that *PARP1* and *BRCA1/BRCA2* are synthetically lethal (Figure 2), as the absence of functional PARP1, along with the deletion of *BRCA1* and *BRCA2,* causes apoptosis (Farmer et al. 2005).

An important observation regarding BRCA is that patients with $BRCA1^{-/-}/BRCA2^{-/-}$ tumour cells have $BRCA1^{+/-}/BRCA2^{+/-}$ somatic cells (Bryant et al., 2005). This implies that the HR pathway of non-cancerous somatic cells will continue to function in the event of PARP1 inhibition. This demonstrates a promising characteristic: PARP1 deletion is not causally linked to oncogenesis. (http://cancer.sanger.ac.uk/cancergenome/assets/cancer_gene_census.xls. Forbes et al., 2008. Accessed August 2012), which may also imply that it is not lethal to somatic cells.

Following the experimental outcomes of the above PARP1 studies the PARP1 inhibitor Olaparib began phase I clinical trials in 2007 (ClinicalTrials.gov, 2012a). The mechanism of action for Olaparib relies on the synthetic lethality of PARP1 inhibition in patients with *BRCA1* or *BRCA2* mutations.

The preliminary results from 2009 were promising and indicated that PARP1 inhibition using Olaparib does not affect patients in the severely adverse ways typical of many chemotherapy treatments. The results also indicated that the anti-cancer activity was restricted to patients exhibiting the *BRCA1* or *BRCA2* mutation, precisely as expected by synthetic lethality experiments (Fong et al. 2009).

Unfortunately, Olaparib was halted prior to phase 3 trials due to less than acceptable measures of patient benefit while being used for the maintenance treatment of serious ovarian cancer (AstraZeneca, 2011). However, it is undergoing a number of phase 2 trials in combination with other chemotherapeutics (ClinicalTrials.gov, 2012b) for its effectiveness in combined therapies. Furthermore, many other PARP1 inhibitors are currently undergoing various phases of clinical trials in addition to combination therapies with classical chemotherapeutics. A search of ClinicalTrials.gov for the key words "PARP1 inhibitor" and "cancer" returned 76 current phased trials as of July 2012.

PTEN, PARP1 and mTOR - An Additional Avenue of Inquiry

Two relatively recent studies into synthetically lethal partners of PARP1 have identified the tumour suppressor PTEN (Mendes-Pereira et al. 2009) and mTOR (Neshat et al. 2009). PTEN is another HR-related protein which implies that sensitivity to PARP1 inhibition may exist in more cancers than those with *BRCA* mutations.

A mutated *PTEN* is found in many cancers, including endometrial cancers, glioblastomas, malignant melanomas, prostate, breast, lung and colorectal cancers (Song et al., 2011). *PTEN* shows synthetic lethality with mTOR. The mTOR protein is not involved in HR; instead, it is involved with the phosphatidyl inositol 3-kinase (PI3-kinase)/Akt cellular growth signalling pathway (Neshat et al. 2001).

This increased sensitivity to mTOR inhibition is apparently due to a constitutive increase in cellular concentrations of phosphatidylinositol (3,4,5) triphosphate (PIP3) in cells without PTEN.

The synthetically lethal effects are best described by recalling oncogenic addiction: the elevated concentration of PIP3 causes the oncogenic cell to be addicted to PI3-kinase/Akt pathway. Inhibiting mTOR, using the rapamycin derivative CCI-779, disables the PI3-kinase/Akt pathway and, because of oncogenic addiction, kills the tumour cell.

The apparent success of PARP1 inhibitors coupled with additional synthetically lethal drug targets would tend to indicate that synthetic lethality could be a new direction in anti-cancer drug development (Iglehart and Silver, 2009). Large scale investigations into synthetically lethal interactions may provide insight into many anti-cancer drug targets.

MOVING BEYOND INDIVIDUAL INTERACTIONS - LARGE SCALE ANALYSIS FOR SYNTHETIC LETHALITY

Gene Centered Synthetic Lethality in Human Cancers

Two recent studies have used RNAi to determine the synthetic lethal partners of the *RAS* gene family member *KRAS* (Barbie et al., 2009; Scholl et al., 2009). These studies knocked out *KRAS* then tested the cells' fitness following shRNA knockdown of selected genes. The studies resulted in only two statistically significant synthetically lethal partners for *KRAS*, the serine/threonine kinase STK33 (Scholl et al., 2009) and the transcriptional control protein NF-κB activator TBK1 (Barbie et al., 2009). Although two synthetically lethal partners of *KRAS* are certainly better than none, these methods are not sufficient when large scale putative anti-cancer drug target predictions are desired. Below we will discuss a recent method based on gene homology that attempts to increase the throughput for discovery of human synthetic lethal interactions using collated synthetic lethality data for *S. cerevisiae*.

Using Homology to Predict Anti-Cancer Gene Targets

A recent study (Conde-Pueyo et al., 2009) using synthetic lethality data from over 1233 articles, collated and regularly updated by the online database "The BioGrid (http://www.thebiogrid.org)" (Stark et al., 2006. Accessed August 2012), created a snapshot of the budding yeast *S. cerevisiae* synthetic lethal network with the intent of finding anti-cancer drug targets *en masse*. This resulted in a gene interaction network with 2383 genes and 12707 synthetically lethal interactions. Interestingly, more than half of the interactions came from five large scale synthetic lethality experiments, indicating that *S. cerevisiae* is an excellent model organism when attempting high-volume synthetic lethal analysis. Finding synthetically lethal interactions at this scale in human cancers would give an enormous boost to the number of potential anti-cancer drug targets. Unfortunately, large-scale, multi-gene knockout synthetic lethal screenings are performed almost exclusively with budding yeast. This is due to the fact that the techniques for high throughput synthetic genetic interaction analysis were only relatively recently devised, and specific to *S. cerevisiae* (Tong and Boone, 2005). This is precisely the obstacle addressed by Conde-Pueyo et al. (2009) with their attempt to use known homologies between yeast and human genes to construct an inferred human synthetic lethal network.

To infer the human synthetic lethal network, the synthetic lethal network for yeast was first constructed from the data source mentioned above. To correlate human and yeast genes, data from the ENSEMBL whole-genome multiple alignment database was used. Due to duplication events in both species post divergence, they sorted gene homology into four types, although only three were used: one-to-one, a single yeast ortholog of a human gene; n-to-one, one human gene has more than one yeast ortholog; one-to-m, one yeast gene has more than one human ortholog; and, n-to-m, a paralogous set of human genes are orthologous to a paralogous set of yeast genes. N-to-m, the 4th set, was not used due to its complexity.

One-to-one homologies were the simplest to use when constructing the inferred human network as the homologous interactions were simply assigned to orthologous genes. In the case of one-to-m, paralogous human genes are combined into a single representation which was then assigned the interactions of the yeast ortholog. Conversely, n-to-one situations were resolved by assigning all the interactions of the yeast paralogs to the single human ortholog. These are visualized in Figure 3.

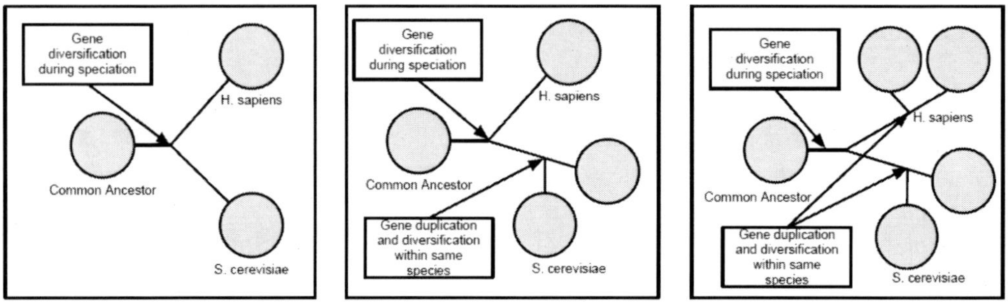

Figure 3. Orthologous relationships considered when constructing the inferred human synthetic lethal network. Figure modified from Conde-Pueyo et al. (2009). The first panel displays a one-to-one relationship of an ortholog between *S. cerevisiae* and *H. sapiens*. Following divergence from the common ancestor, each gene followed a single evolutionary trajectory in each species. The second panel displays a relationship between a single ortholog in *H. sapiens* and a set of orthologous paralogs in *S. cerevisiae* (n-to-one). The final panel displays a mapping between paralogs in *H. sapiens* that are orthologous to paralogs in *S. cerevisiae* (m-to-n).

The results from the inference found that 52% of the 2383 genes in the yeast synthetic lethal network have orthologs in humans. However, only 45% had a synthetically lethal partner, the other 7% of genes were without a synthetically lethal partner after inference and removed. This resulted in 1002 synthetically lethal human orthologs. Interestingly, the majority of those genes involved in synthetically lethal interactions (933) formed a giant connected component.

A giant connected component is a large set of interactions where any single gene in the component can be traced to another gene in the component via some path of gene intermediaries. Using the Gene ontology database, it was noted that nearly half of these genes (49%) have biological functions attributed to proteasomal protein catabolic processes (24%) or Golgi vesicle transport (25%). The remaining major groups accounted for another 40.8%, and were composed of genes related to DNA replication (6.8%), DNA repair (7.4%), response to DNA damage stimulus (8.1%), RNA processing (8.7%), and cell cycle regulation (9.8%).

Following the construction of the inferred network, the COSMIC database was used to identify cancer related genes in the network. This ultimately resulted in a reduction of the network down to 59 cancer related genes with 694 interactions with 347 synthetically lethal partners. The homology approach does have certain limitations; for example, it will not find non-homologous synthetically lethal partners. Another limitation of the homology inference approach, as noted by Conde-Pueyo et al. (2009), is that while there is a strong conservation of orthologous protein function, network interactions may not completely correlate between species. For example, approximately 29% of synthetically lethal interactions are conserved between *S. cerevisiae* and *Schizosaccharomyces pombe* (Dixon et al., 2008). Only 0.7% of interactions were conserved between the more distantly related species *S. cerevisiae* and *Caenorhabditis elegans* (Tischler et al., 2008).

Despite these limitations, the homology approach as described in this review does provide an interesting avenue for drug target discovery.

The major contribution is the 347 possible drug targets deduced from a single study, much larger than those described when utilizing human cells, providing many avenues of inquiry for novel drug target discovery.

Conclusion

In this chapter, we have discussed a relatively recent approach to anti-cancer drug development using synthetic lethality. We introduced synthetic lethality as an already clinically useful method of drug discovery with the PARP1 inhibitor Olaparib. We described how *BRCA1* or *BRCA2* mutations can contribute to PARP1 inhibition sensitivity to synthetic lethality. In addition, the interaction of PARP1 and PTEN was described for cancers that do not necessarily exhibit *BRCA* mutations. We further discussed the PTEN/mTOR interaction and how its inhibition resulted in synthetic lethality when *PTEN* contained a deletion.

On the topic of large scale studies, we briefly discussed two RNAi experiments for $KRAS^{-/-}$ mutants that revealed two synthetically lethal partners serine/threonine kinase STK33 and NF-κB. These may be possible anti-cancer drug targets.

We finished by discussing a novel approach to synthetic lethality in human cells using homology with yeast genes. This method identified 59 orthologs in human cells that interact with 347 partners in a synthetically lethal manner, all possible drug targets.

The search for targeted anti-cancer therapeutics is an ongoing process. With the volume of genetic and gene interaction data being generated every year, advances will likely be made in their discovery. For example, large scale synthetic gene array (SGA) studies have been conducted for *S. cerevisiae* (Costanzo et al., 2010).

Additionally, this technology was recently extended to the fission yeast *S. pombe* (Baryshnikova et al., 2010). This technology allows the use of double knockouts to determine their effect on phenotype on a very large scale (5.7 million gene pairs in *S. cerevisiae*)(Costanzo et al., 2010). Further extension may allow the technology to be used for human cells.

Based on the advances discussed in this paper, the next decade will likely prove very exciting for our understanding of human cancer genetics, and particularly how we utilize oncogene addiction and synthetic lethality to develop new therapeutics.

ACKNOWLEDGMENTS

We gratefully acknowledge the support by the University of Victoria Faculty Research Grant #10000-71117 and Natural Sciences and Engineering Research Council Discovery Grant #138216-2009 to F.Y.M. Choy.

REFERENCES

Ashworth, A., Lord, C.J., and Reis-Filho, J.S. (2011). Genetic interactions in cancer progression and treatment. *Cell 145*, 30-38.

AstraZeneca (2011). AstraZeneca updates on olaparib and TC-5214 development programmes.

Barbie, D.A., Tamayo, P., Boehm, J.S., Kim, S.Y., Moody, S.E., Dunn, I.F., Schinzel, A.C., Sandy, P., Meylan, E., Scholl, C., et al. (2009). Systematic RNA interference reveals that oncogenic KRAS-driven cancers require TBK1. *Nature 462*, 108-112.

Baryshnikova, A., Costanzo, M., Dixon, S., Vizeacoumar, F.J., Myers, C.L., Andrews, B., and Boone, C. (2010). Synthetic genetic array (SGA) analysis in Saccharomyces cerevisiae and Schizosaccharomyces pombe. *Methods Enzymol. 470*, 145-179.

Berns, K., Hijmans, E.M., Mullenders, J., Brummelkamp, T.R., Velds, A., Heimerikx, M., Kerkhoven, R.M., Madiredjo, M., Nijkamp, W., Weigelt, B., et al. (2004). A large-scale RNAi screen in human cells identifies new components of the p53 pathway. *Nature 428*, 431-437.

Birchler, J.A., Riddle, N.C., Auger, D.L., and Veitia, R.A. (2005). Dosage balance in gene regulation: biological implications. *Trends Genet. 21*, 219-226.

Birchler, J.A., and Veitia, R.A. (2007). The gene balance hypothesis: from classical genetics to modern genomics. *Plant Cell 19*, 395-402.

Birchler, J.A., and Veitia, R.A. (2010). The gene balance hypothesis: implications for gene regulation, quantitative traits and evolution. *New Phytol. 186*, 54-62.

Bryant, H.E., Schultz, N., Thomas, H.D., Parker, K.M., Flower, D., Lopez, E., Kyle, S., Meuth, M., Curtin, N.J., and Helleday, T. (2005). Specific killing of BRCA2-deficient tumours with inhibitors of poly (ADP-ribose) polymerase. *Nature 434*, 913-917.

ClinicalTrials.gov (2012a). Search results for olaparib. http://clinicaltrials.gov/ct2/show /NCT01078662

ClinicalTrials.gov (2012b). A Study to Assess the Safety and Pharmacokinetics of an Inibitor of Poly ADP-Ribose Polymerase-1 (PARP1). http://clinicaltrials.gov/ct2/show /NCT00516373.

Conde-Pueyo, N., Munteanu, A., Solé, R.V., and Rodríguez-Caso, C. (2009). Human synthetic lethal inference as potential anti-cancer target gene detection. *BMC Syst. Biol. 3*, 116.

Conde, C., Mark, M., Oliver, F.J., Huber, A., De Murcia, G., and Ménissier-de Murcia, J. (2001). Loss of poly (ADP-ribose) polymerase-1 causes increased tumour latency in p53-deficient mice. *EMBO J. 20*, 3535-3543.

Costanzo, M., Baryshnikova, A., Bellay, J., Kim, Y., Spear, E.D., Sevier, C.S., Ding, H., Koh, J.L.Y., Toufighi, K., Mostafavi, S., et al. (2010). The genetic landscape of a cell. *Science's STKE 327*, 425.

Dixon, S.J., Fedyshyn, Y., Koh, J.L.Y., Prasad, T.S.K., Chahwan, C., Chua, G., Toufighi, K., Baryshnlkova, A., Hayles, J., Hoe, K., Kim, D., Park, H., Myers, C.L., Pandy, A., Durocher, D., Andrews, B.J., and Boone, C. (2008). Significant conservation of synthetic lethal genetic interaction networks between distantly related eukaryotes. *PNAS 105*, 16653-16658.

Earhart, R.H., Martin, P.A., Tutsch, K.D., Ertürk, E., Wheeler, R.H., and Bull, F.E. (1983). Improvement in the therapeutic index of cisplatin (NSC 119875) by pharmacologically induced chloruresis in the rat. *Cancer Res. 43*, 1187-1194.

Farmer, H., McCabe, N., Lord, C.J., Tutt, A.N.J., Johnson, D.A., Richardson, T.B., Santarosa, M., Dillon, K.J., Hickson, I., Knights, C., et al. (2005). Targeting the DNA repair defect in BRCA mutant cells as a therapeutic strategy. *Nature 434*, 917-921.

Fong, P.C., Boss, D.S., Yap, T.A., Tutt, A., Wu, P., Mergui-Roelvink, M., Mortimer, P., Swaisland, H., Lau, A., O'Connor, M.J., et al. (2009). Inhibition of poly (ADP-ribose) polymerase in tumors from BRCA mutation carriers. *New Engl. J. Med. 361*, 123-134.

Forbes S.A., Bhamra, G., Bamford, S., Dawson, E., Kok, C., Clements, J., Menzies, A., Teaque, J.W.,

Futreal, P.A., Stratton, M.R. (2008). The Catalogue of Somatic Mutations in Cancer (COSMIC). *Curr Protoc Hum Genet,* Chapter 10: Unit 10, 11.

Futreal, P.A., Coin, L., Marshall, M., Down, T., Hubbard, T., Wooster, R., Rahman, N., and Stratton, M.R. (2004). A census of human cancer genes. *Nat. Rev. Cancer 4*, 177-183.

Hanahan, D., and Weinberg, R.A. (2000). The hallmarks of cancer. *Cell 100*, 57-70.

Hanahan, D., and Weinberg, R.A. (2011). Hallmarks of cancer: the next generation. *Cell 144*, 646-674.

Hart, T.G., Ramani, A.K. and Marcotte, E.M. (2006) How complete are current yeast and human protein-interaction networks?. *Genome Biol.* 7, 120

Iglehart, J.D., and Silver, D.P. (2009). Synthetic lethality—a new direction in cancer-drug development. *New Engl. J. Med. 361*, 189-191.

Kaelin, W.G. (2005). The concept of synthetic lethality in the context of anticancer therapy. *Nat. Rev. Cancer 5*, 689-698.

Kang, S.M., Tran, A.C., Grilli, M., and Lenardo, M.J. (1992). NF-kappa B subunit regulation in nontransformed CD4+ T lymphocytes. *Science 256*, 1452-1456.

Klein, G. (1981). The role of gene dosage and genetic transpositions in carcinogenesis. *Nature 294*, 313-318.

Lando, M., Holden, M., Bergersen, L.C., Svendsrud, D.H., Stokke, T., Sundfør, K., Glad, I.K., Kristensen, G.B., and Lyng, H. (2009). Gene dosage, expression, and ontology analysis identifies driver genes in the carcinogenesis and chemoradioresistance of cervical cancer. *PLoS Genet. 5*, e1000719.

Li, J., Yen, C., Liaw, D., Podsypanina, K., Bose, S., Wang, S.I., Puc, J., Miliaresis, C., Rodgers, L., McCombie, R., et al. (1997). PTEN, a putative protein tyrosine phosphatase gene mutated in human brain, breast, and prostate cancer. *Science 275*, 1943-1947.

Mendes-Pereira, A.M., Martin, S.A., Brough, R., McCarthy, A., Taylor, J.R., Kim, J.S., Waldman, T., Lord, C.J., and Ashworth, A. (2009). Synthetic lethal targeting of PTEN mutant cells with PARP1 inhibitors. *EMBO Mol. Med. 1*, 315-322.

Neshat, M.S., Mellinghoff, I.K., Tran, C., Stiles, B., Thomas, G., Petersen, R., Frost, P., Gibbons, J.J., Wu, H., and Sawyers, C.L. (2001). Enhanced sensitivity of PTEN-deficient tumors to inhibition of FRAP/mTOR. *PNAS 98*, 10314.

Nijman, S.M.B. (2011). Synthetic lethality: general principles, utility and detection using genetic screens in human cells. *FEBS Lett. 585*, 1-6.

Olson, R.D., and Mushlin, P.S. (1990). Doxorubicin cardiotoxicity: analysis of prevailing hypotheses. *The FASEB J. 4*, 3076-3086.

Rottensteiner, H., Kal, A.J., Hamilton, B., Ruis, H., and Tabak, H.F. (1997). A heterodimer of the Zn2Cys6 transcription factors Pip2p and Oaf1p controls induction of genes encoding peroxisomal proteins in Saccharomyces cerevisiae. *Eur. J. Biochem. 247*, 776-783.

Rudders, R.A., and Hensley, G.T. (1973). Bleomycin pulmonary toxicity. *Chest 63*, 626-628.

Scholl, C., Fröhling, S., Dunn, I.F., Schinzel, A.C., Barbie, D.A., Kim, S.Y., Silver, S.J., Tamayo, P., Wadlow, R.C., Ramaswamy, S., et al. (2009). Synthetic Lethal Interaction between Oncogenic KRAS Dependency and STK33 Suppression in Human Cancer Cells. *Cell 137*, 821-834.

Song, M.S., Carracedo, A., Salmena, L., Song, S.J., Egia, A., Malumbres, M., and Pandolfi, P.P. (2011). Nuclear PTEN Regulates the APC-CDH1 Tumor-Suppressive Complex in a Phosphatase-Independent Manner. *Cell 144*, 187-199.

Stark, C., Breitkreutz, B.J., Reguly, T., Boucher, L., Breitkreutz, A., and Tyers, M. (2006). BioGRID: a general repository for interaction datasets. *Nucleic Acids Res. 34*, (Database issue):D535-9.

Stentoft, J. (1990). The toxicity of cytarabine. *Drug Safety 5*, 7.

Tischler, J.,Lehner, B., and Fraser, A.G. (2008) Evolutionary plasticity of genetic interaction networks. *Nature Genet.* 40, 390-391.

Tong, A.H.Y., and Boone, C. (2005). Synthetic genetic array analysis in Saccharomyces cerevisiae. *Methods Mol. Biol. 313*, 171-192.

Weinstein, I.B. (2002). Addiction to oncogenes-the Achilles heal of cancer. *Science 297*, 63-64.

Weinstein, I.B., and Joe, A. (2008). Oncogene addiction. *Cancer Res. 68*, 3077-3080.

BRIEF REPORT

In: Biochemistry and Molecular Biology Coming of Age
Editor: Mary Miu Yee Waye
ISBN: 978-1-63321-815-4

Chapter 8

Sanfilippo Syndrome: Identification and Removal of a Cryptic Splice Site in Alpha-N-Acetylglucosaminidase cDNA for Potential Improved Expression and Enzyme Replacement Therapy

Sarah N. Truelson, R. Rebecca Jantzen and Francis Y. M. Choy*
Department of Biology, Biomedical Research Centre, University of Victoria
Victoria, British Columbia, Canada

Abstract

α-N-Acetylglucosaminidase (Naglu) is a lysosomal acid hydrolase that is deficient in patients with Mucopolysaccharidosis III type B (MPS IIIB). Naglu deficiency causes cytotoxic accumulation of heparan sulfate in lysosomes; this is especially detrimental to the central nervous system, resulting in neurodegeneration, behavioural problems, mental retardation, and shortened lifespan. Our laboratory is currently working to express Naglu fused to a synthetic protein transduction domain with the ultimate goal of developing an efficient system to produce human recombinant Naglu for enzyme replacement therapy. Conjugation of Naglu to a protein transduction domain may allow the enzyme to cross the blood-brain barrier to treat the neurological symptoms of MPS IIIB.

The Naglu fusion protein has been successfully produced in an active state in recombinant *Spodoptera frugiperda* 9 (*Sf*9) cultures; however, low resultant expression levels have prevented further characterization studies. Degradation of Naglu mRNA due to activation of a cryptic splice site in human Naglu cDNA was hypothesized, as we consistently detected the presence of a small, non-target DNA band in reverse transcription polymerase chain reaction (RT-PCR) experiments. Sequence analysis showed that the 426 bp band contained Naglu sequence corresponding to the 5' and 3' ends of the gene, with a large central portion missing, and was likely the result of cryptic

* Corresponding author: F.Y.M. Choy, Department of Biology, Biomedical Research Centre, University of Victoria, 3800, Finnerty Road, Victoria, BC, Canada V8P 5C2. Tel. (250)721-7107; FAX: (250)721-7120; E-mail: fchoy@uvic.ca.

splicing in the *Sf9* expression system. Site-directed mutagenesis by fusion PCR was used to alter the splice donor and acceptor sites in Naglu cDNA; RT-PCR on RNA isolated from *Sf9* cells transfected with the new construct indicated presence of the full length 2.2 Kb Naglu product and absence of the 426 bp band. Use of this construct, which successfully eliminated cryptic splicing, could potentially increase recombinant Naglu expression in *Sf9* cells.

Keywords: Sanfilippo syndrome, α-N-acetylglucosaminidase, cryptic splice site, site-directed mutagenesis, *Sf9* cell expression system

INTRODUCTION

Mucopolysaccharidosis III (MPS III), also known as Sanfilippo syndrome, is a panethnic autosomal recessive metabolic disorder comprising four distinct biochemical subtypes and occurring at an incidence of approximately one in 24,000 live births (van de Kamp et al., 1981). Mucopolysaccharidosis III type B (MPS IIIB) is caused by deficiency of the lysosomal acid hydrolase α-N-acetylglucosaminidase (Naglu). Naglu catalyzes an intermediate step in heparan sulfate catabolism by hydrolyzing the α-(1,4)-glycosidic bond between N-acetylglucosamine and uronic acid. In the absence of functional Naglu, accumulation of partially degraded heparan sulfate in lysosomes results in neurotoxicity causing progressive cerebral atrophy (Zhao et al., 1998). MPS IIIB shows an extremely wide clinical heterogeneity, with patients typically demonstrating mild somatic symptoms but profound central nervous system involvement. The resultant neural degeneration causes developmental delays, aggressive behaviour, mental retardation, and a significantly shortened lifespan (Schmidtchen et al., 1998). Enzyme replacement therapy is currently ineffective due to the inability of exogenous Naglu to cross the blood-brain barrier.

Our laboratory is currently using a *Spodoptera frugiperda* (*Sf9*) insect cell system to express Naglu fused to an eleven amino acid protein transduction domain, identified in the TAT protein of the HIV virus, that has been shown to transport its fusion partners across the blood-brain barrier *in vivo* (Schwarze et al., 1999). Our aim is to develop an efficient system to produce human recombinant Naglu for enzyme replacement therapy. High levels of active human recombinant enzymes have been produced in *Sf9* cells (Pfeifer et al., 2001), including lysosomal glucocerebrosidase, which was secreted at concentrations of 10 to 15 mg/L (Sinclair et al., 2006). However, we have consistently experienced lower than expected expression levels of Naglu. Using various Naglu constructs, Bandsmer (2004) observed a lower average level of secretion around 175 ng/mL.

Reverse transcription polymerase chain reaction (RT-PCR) using RNA isolated from *Sf9* cells expressing recombinant human Naglu fused to the protein transduction domain (modTAT•Naglu) revealed the presence of a non-target band about 400 bp in length, which is much smaller than full length Naglu at 2.2 Kb. Sequence data indicated that this 426 bp fragment contained sequence from both the 5' and 3' ends of Naglu cDNA with a large central portion missing; the internally deleted segment of 1735 bp in length started with a GU dinucleotide and ended with an AG dinucleotide – the most common splice donor and acceptor signals, respectively (Burset et al., 2000).

We hypothesize that a cryptic splice site in human Naglu cDNA may be causing degradation of Naglu mRNA when expressed in *Sf9* cells, contributing to a low yield of active recombinant Naglu. In this communication, we describe a procedure that utilizes site directed mutagenesis by fusion polymerase chain reaction (PCR) to modify donor and acceptor splice signals without altering the amino acid sequence. Use of this procedure created a construct that successfully eliminated Naglu splicing in *Sf9* expression cultures.

MATERIALS AND METHODS

Cultivation of *Sf9* Cells for DNA and RNA Extraction

Sf9 cultures containing the modTAT•Naglu expression plasmid were prepared as previously described by Bandsmer et al. (2006). *Sf9* modTAT•Naglu cells were cultured according to Invitrogen protocols prior to total RNA extraction using the Qiagen RNeasy® Mini Kit and Qiashredder® column. RNeasy® samples were deoxyribonuclease treated using the Ambion TURBO DNA-free™ kit according to the manufacturer's instructions. Genomic DNA was purified from a confluent modTAT•Naglu culture using the Qiagen DNeasy® Tissue Kit; recombinant modTAT•Naglu *Sf9* cultures have the expression vector integrated into genomic DNA.

SuperScript™ II RT-PCR and Sequence Analysis

Reverse transcription (RT) was performed using 80 ng of total RNA according to the Invitrogen SuperScript™ II RT protocol. PCR was performed on RT samples and a genomic modTAT•Naglu culture DNA sample. PCR reactions contained the following reagents: 1 μL RT template or 20 ng genomic DNA template, 0.25 mM deoxynucleotide triphosphates (dNTPs), 1X PCR Buffer minus Mg, 1.5 mM $MgCl_2$, 8% v/v dimethyl sulfoxide (DMSO), 1.3 M betaine, 1 U Platinum® Taq DNA polymerase and 0.6 μM of each primer (forward: 5'CGACGAGGCCCGGGAGG3', reverse: 5'CCAAGAGCCGGCCACCCAG3') in a final volume of 50 μL. Samples underwent initial denaturation at 94°C for 5 min, followed by 10 cycles of 94°C for 1.5 min, 64°C for 1 min, and 72°C for 1.75 min, 25 cycles of 94°C for 1.5 min, 62°C for 1 min, and 72°C for 1.75 min, and final elongation at 72°C for 7 min. Following agarose gel electrophoresis and ethidium bromide staining, the band of just over 400 bp in size was excised and purified using the Qiagen QIAquick™ Gel Extraction Kit. DNA sequence analysis of the PCR product was performed using a capillary-based sequencer (Beckman Coulter CEQ™ 8000 Genetic Analysis System) and the CEQ Dye Terminator Cycle Sequencing Kit (Beckman Coulter) at the Centre for Biomedical Research DNA Sequencing Facility (University of Victoria) using the primers listed above.

C.therm. One-Step RT-PCR

The *C.therm.* Polymerase One-step RT-PCR System (Roche) was used with RNA from a modTAT•Naglu culture, as the *Carboxydothermus hydrogenoformans* (*C.therm.*) bacterial polymerase is able to eliminate RT-PCR artifacts resulting from viral reverse transcriptase template switching. The reaction mixture contained 80 ng of total RNA, 0.4 mM dNTPs, 1X *C.therm.* buffer, 10% v/v DMSO, 1.3 M betaine, 30 mM dithiothreitol, 20 U RNaseOUT™ Ribonuclease Inhibitor, 2 μL *C.therm.* polymerase mixture and 0.3 μM of each primer (same as in SuperScript™ II RT-PCR) in a final volume of 50 μL. RT-PCR amplification proceeded as follows: RT at 60°C for 30 min, denaturation at 94°C for 2 min, 35 cycles of 94°C for 1 min, 60°C for 1 min and 72°C for 2 min, and final elongation at 72°C for 7 min.

Site-Directed Mutagenesis Fusion PCR to Remove Cryptic Splice Site

Table 1 indicates the six primers used for site-directed mutagenesis of the modTAT•Naglu cDNA construct. The primers allowed for amplification of three fragments for use in a single fusion PCR reaction. Primers internal to the Naglu cDNA sequence contained mismatched bases in order to modify splice donor and acceptor signals (Figure 1).

Table 1. Primers used in site-directed mutagenesis to create mutagenized Naglu construct

Primer	Sequence (5'→3')	Description
A	A*GAATTC*GACGAGGCCCGG GAGG	Forward primer corresponding to bases 1-15 of Naglu cDNA construct, EcoRI site shown in italics
B	**CCACGCCCAACCGTTACC** **GCTATTACC**AGAATGTG	Forward primer corresponding to bases 302-336, bold indicates overlap with primer D, mismatched bases underlined
C	**CAACTGGAGCAAGCCTTC** **GTTCGC**AGCAAG	Forward primer corresponding to bases 2038-2067, bold indicates overlap with primer E, mismatched base underlined
D	**GTAATAGCGGTAACGGTT** **GGGCGTGG**CCTC	Reverse primer corresponding to bases 298-327, bold indicates overlap with primer B, mismatched bases underlined
E	**GAGAACGAAGGCTTGCTC** **CAGTTG**GAAGAC	Reverse primer corresponding to bases 2032-2061, bold indicates overlap with primer C, mismatched base underlined
F	A*TCTAGA*TTGCACGTGCCT GACG	Reverse primer corresponding to bases 2184-2198, XbaI site shown in italics (includes stop codon)

Note – Mutagenesis was implemented using fusion PCR to alter the cryptic donor and acceptor splice sites in modTAT•Naglu cDNA. Base locations refer to positions within the modTAT•Naglu construct created by Judith Bandsmer (Bandsmer et al., 2006), where base 1 starts at the first nucleotide of Naglu cDNA and corresponds to amino acid 24 of reference Naglu (NCBI accession NM_000263.3).

Mutagenesis at donor splice signal:

```
N:      707    GAG  GCC  ACG  CCC  AAC  AG|G  TAC  CGC  TAT  TAC  CAG  AAT  GTG
745
M:      298    GAG  GCC  ACG  CCC  AAC  CG T  TAC  CGC  TAT  TAC  CAG  AAT  GTG
336

P:             Glu  Ala  Thr  Pro  Asn  Arg   Tyr  Arg  Tyr  Tyr  Gln  Asn  Val
```

Mutagenesis at acceptor splice signal:

```
N:      2441   GTC  TTC  CAA  CTG  GAG  CAG|  GCC  TTC  GTT  CTC  AGC  AAG   2476
M:      2032   GTC  TTC  CAA  CTG  GAG  CAA   GCC  TTC  GTT  CTC  AGC  AAG   2067

P:             Val  Phe  Gln  Leu  Glu  Gln   Ala  Phe  Val  Leu  Ser  Lys
```

Figure 1. Modification of donor and acceptor splice signals in Naglu cDNA sequence. N shows the sequence for reference human Naglu cDNA (NCBI accession NM_000263.3), while M indicates the post-mutagenesis cDNA sequence. Location of splicing in the original construct is indicated by |. GT and AG dinucleotides are italicized. P shows the amino acid sequence, which remains unchanged following mutagenesis. Nucleotide 1 of the laboratory Naglu construct corresponds to nucleotide 410 of NM_000263.3 reference sequence.

The first 343 bp fragment that covered the 5' portion of Naglu was amplified in the following reaction mixture: 10 ng modTAT•Naglu plasmid template, 0.25 mM dNTPs, 1X High Fidelity PCR Buffer, 1.5 mM $MgSO_4$, 8% v/v DMSO, 1.3 M betaine, 1 U Platinum® Taq DNA Polymerase High Fidelity, and 0.6 μM each of primers A and D in a total volume of 50 μL. The reaction was run as follows: denaturation at 94°C for 5 min, 35 cycles of 94°C for 30 s, 61°C for 30 s and 72°C for 30 s, and final elongation at 72°C for 5 min. The fragment was purified using the QIAquick™ Gel Extraction Kit.

The second and third fragments that covered the middle and 3' portions of modTAT•Naglu, of 1760 bp and 169 bp in size, respectively, were amplified in the following reactions: 10 ng modTAT•Naglu plasmid template, 0.25 mM dNTPs, 1X High Fidelity PCR Buffer, 1.5 mM $MgSO_4$, 1 U Platinum® Taq DNA Polymerase High Fidelity, and 0.6 μM each of primers B and E for fragment two or C and F for fragment three, in a total volume of 50 μL.

Reactions were run as follows: initial denaturation at 94°C for 5 min, 35 cycles of 94°C for 30 s, 56°C for 30 s and 72°C for 2 min, and final elongation at 72°C for 7 min. Fragments were purified using the QIAquick™ Gel Extraction Kit.

The fusion PCR reaction mixture contained 75 ng of each fragment, 0.25 mM dNTPs, 1X High Fidelity PCR Buffer, 1.5 mM $MgSO_4$ and 1 U Platinum® Taq DNA Polymerase High Fidelity in a total volume of 50 μL. Negative control reactions contained 75 ng of each fragment individually to ensure that gel extractions did not contain any contaminating modTAT•Naglu plasmid template.

An initial denaturation step at 94°C for 5 min was followed by seven cycles of 94°C for 15 s and 68°C for 2 min. To amplify the complete mutagenesis fusion product, primers A and F were added to a final concentration of 0.3 μM and reactions proceeded as follows: 10 cycles of 94°C for 15 s, 64°C for 30 s and 68°C for 2 min, 15 cycles of 94°C for 15 s, 62°C for 30 s and 68°C for 1 min, and final elongation at 68°C for 7 min. The fusion amplicon was excised and purified using the QIAquick™ Gel Extraction Kit.

Construction of mut-mod•TAT Plasmid and Transfection of *Sf9* Cells

Mutagenized fusion product was cloned using the Promega pGEM®-T vector system and *E. coli* JM109 chemically competent cells according to the manufacturer's protocol. Plasmid DNA from a colony containing the cloned 2.2 Kb mutagenized fusion product was digested with EcoRI and XbaI in preparation for ligation into the expression vector. The original modTAT•Naglu expression plasmid was also digested with EcoRI and XbaI in order to prepare doubly digested vector.

Digested insert was ligated into the digested expression vector using Invitrogen T4 DNA Ligase. *E. coli* JM109 cells were transformed with the ligation mixture and selected on low salt Luria-Bertani agar containing 25 μg/mL Zeocin™. Colonies were screened using PCR with mutagenesis primers B and E as described in the protocol to produce the 1760 bp fusion fragment. Plasmid DNA was isolated from colonies containing the mutagenized modTAT•Naglu (mut-modTAT•Naglu) construct and the insert was sequenced as described previously.

Plasmid DNA containing the sequence-confirmed mutagenesis construct was used to transfect Invitrogen SFM adapted *Sf9* cells using Cellfectin® II reagent according to the manufacturer's protocol.

Transfected cells were selected with increasing concentrations of Zeocin™ (500 μg/mL to 1 mg/mL) to create stable polyclonal cell lines with the incorporated mut-modTAT•Naglu expression vector.

RT-PCR Analysis of mut-modTAT•Naglu *Sf9* Cell Lines

Total RNA was isolated from a mut-modTAT•Naglu culture using the RNeasy® Mini Kit. RNA samples were reverse transcribed according to the previous SuperScript™ II RT protocol. PCR was then used to amplify Naglu cDNA from a modTAT•Naglu culture (used in previous SuperScript™ II RT-PCR) and a mut-modTAT•Naglu culture.

Reactions contained the following: 1 μL RT template, 0.25 mM dNTPs, 1X PCR Buffer minus Mg, 1.5 mM $MgCl_2$, 8% v/v DMSO, 1.3 M betaine, 1 U Platinum® Taq DNA polymerase and 0.6 μM of each primer (same as in original SuperScript™ II RT-PCR) in a final volume of 50 μL. Reactions mixtures were cycled as follows: initial denaturation at 94°C for 5 min, 35 cycles of 94°C for 1 min, 62°C for 1 min and 72°C for 1.75 min, and final elongation at 72°C for 7 min.

RESULTS

RT-PCR Identification of Putative Splice Product

RT-PCR of RNA isolated from a stable modTAT•Naglu *Sf9* culture (Figure 2) produced a 2.2 Kb band as is expected for full-length Naglu. An intensely stained band near 400 bp was also consistently amplified from modTAT•Naglu cDNA samples; this band was not present upon PCR amplification of genomic DNA from a modTAT•Naglu culture.

Sequence analysis revealed that the amplicon was 426 bp in length and contained 100% sequence identity to 5' and 3' regions of Naglu cDNA (matching positions 410-723 and 2459-2548 of reference sequence NM_000263.3). The amplicon consisted of sequence up until a GT site in the Naglu cDNA sequence and continued following an AG site. Comparison to the Naglu cDNA sequence also indicated a 4 bp direct repeat of nucleotide sequence CAGG at deletion junctions.

Results of RT-PCR Using *C.therm.* Polymerase

The *C.therm.* RT-PCR was used to eliminate the possible occurrence of template switching during RT, which can create aberrant deletion-containing RNA signals. The RT-PCR was able to generate the 426 bp putative splice product (data not included), thus eliminating the possibility that the 426 bp band was an artifact generated by template switching.

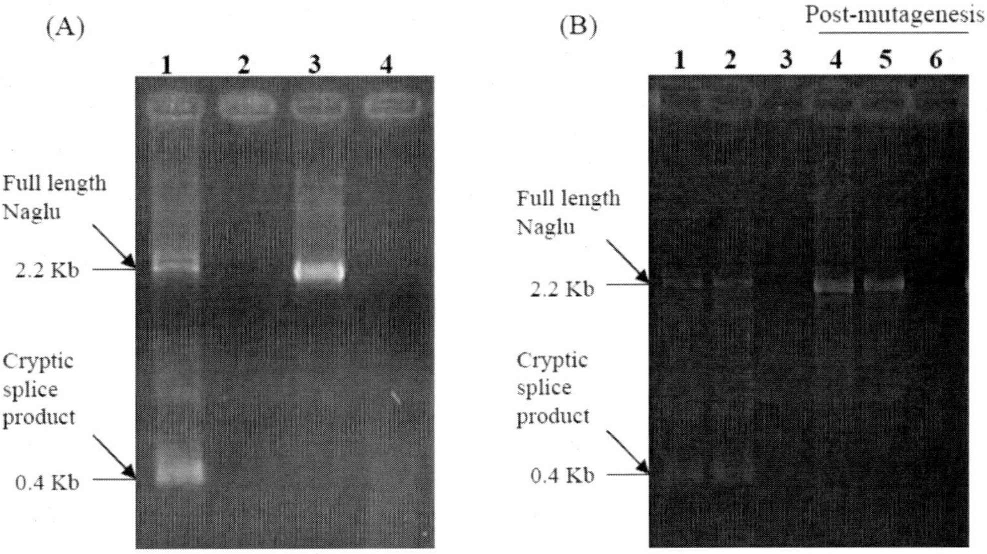

Figure 2. RT-PCR indicates the presence of a truncated Naglu mRNA species and its subsequent elimination upon mutagenesis of cryptic splice donor and acceptor sites. (A) RT-PCR results for modTAT•Naglu culture RNA with modTAT•Naglu genomic DNA as PCR positive control. Presence of a small band near 0.4 Kb is suggestive of RNA degradation; sequence analysis supports that the product was generated through splicing of Naglu mRNA. Lane 1, RT-PCR using modTAT•Naglu culture RNA. Lane 2, RT-PCR negative control: no reverse transcriptase added. Lane 3, PCR positive control using *Sf9* modTAT•Naglu genomic DNA. Lane 4, PCR negative control: no template in reaction. (B) RT-PCR for modTAT•Naglu and mut-modTAT•Naglu culture RNA indicates elimination of splicing upon mutagenesis. Lanes 1 and 2, RT-PCR using modTAT•Naglu culture RNA. Lane 3, RT-PCR negative control for modTAT•Naglu RNA: no reverse transcriptase added. Lanes 4 and 5, RT-PCR using mut-modTAT•Naglu culture RNA. Lane 6, RT-PCR negative control for mut-modTAT•Naglu RNA: no reverse transcriptase added.

Elimination of Splicing Through Site-Directed Mutagenesis

Site-directed mutagenesis by fusion PCR was used to remove donor and acceptor splice signals in the modTAT•Naglu cDNA sequence. Fusion PCR through combination of three mutagenized Naglu fragments resulted in successful production of the full length 2.2 Kb mutagenesis product. Upon cloning into the final expression vector, sequence analysis of the construct indicated presence of the altered bases and no further mutations in the Naglu construct. *Sf9* cells were transfected with the mut-modTAT•Naglu construct and stable cell lines were created using Zeocin™ selection. RT-PCR of RNA isolated from a polyclonal mut-modTAT•Naglu *Sf9* culture resulted in production of the expected 2.2 Kb product for full length Naglu and elimination of the 426 bp putative splice band.

Discussion

RT-PCRs with RNA extracted from *Sf9* cultures expressing the modTAT•Naglu construct indicated a 426 bp band that was much smaller than the 2.2 Kb product expected for full length Naglu. Sequence data from the 426 bp band showed that it was not a spurious PCR product; rather, the amplicon contained sequence corresponding to the 5' and 3' ends of the expected Naglu cDNA sequence, with an internally deleted region 1735 bp in length. No 426 bp product was obtained from PCR amplification of *Sf9* modTAT•Naglu genomic DNA, which contained the stably integrated expression vector, thus the small product was not an artifact of the PCR step. A possibility remained that the band could have been an artifact of the RT step. Of note was a 4 bp direct repeat at deletion junctions; such precise, reproducible deletions between direct repeats have been shown to result from the template switching activity of viral reverse transcriptases (Bowman et al., 1998; Delviks and Pathak, 1999; Zhang and Temin, 1994). Intramolecular template switching occurs most frequently between homologous sequences and when facilitated by a strong RNA secondary structure, and Naglu's GC content of around 64% could contribute to the formation of such a structure (Chan et al., 2009). The *C.therm.* RT-PCR system was implemented as the bacterial *C.therm.* DNA polymerase Klenow fragment, which has residual reverse transcriptase activity and presumably does not possess the inherent template switching activity of viral reverse transcriptases, has been shown to eliminate generation of erroneous RNA signals due to template switching (Mader et al., 2001). The higher temperature at which the RT step was performed and presence of betaine and DMSO in the reaction would have also had the effect of reducing nucleic acid secondary structure (Henke et al., 1997; Winship, 1989). Presence of the putative splice product in the *C.therm.* RT-PCR provided further evidence that the 426 bp band arose from a truly extant RNA splice product rather than from reverse transcriptase template switching at deletion junctions.

Most likely, the small product resulted from a cryptic splice site in the Naglu cDNA sequence that caused Naglu mRNA to be degraded by the *Sf9* spliceosome. In nature, mutations that disrupt an active splice site can result in activation of a nearby cryptic site (Kralovicova et al., 2005; Nelson and Green, 1990). Elimination of naturally active splice sites in cDNA constructs can also result in cryptic splice site activation in the laboratory setting. Bukovac et al. (2008) recognized a cryptic splice site in the cDNA for glucocerebrosidase that resulted in low yields of the enzyme from transfected Chinese hamster ovary cells. Alteration of the cryptic splice signals to eliminate splicing resulted in a six-fold increase in recombinant glucocerebrosidase expression.

In the modTAT•Naglu cryptic splice product, the deleted segment began with a GU dinucleotide and ended with an AG dinucleotide, which are the most common donor and acceptor recognition sites, accounting for approximately 99% of splicing (Burset et al., 2000). Since not all GU and AG dinucleotides within an RNA sequence are involved in splicing, sequence context is also an important factor in determining whether or not splicing will occur. Mount (1982) reported the following consensus sequences for classical GU-AG spliceosomal splicing: the donor sequence is (A or C)AG|GT(A or G)AGT, while the acceptor site contains a pyrimidine stretch followed by N(C or T)AG|G, where | indicates the splice site and N is any nucleotide. The Naglu cDNA cryptic splice signals conform to the core portions of these consensus sequences, with a donor sequence of CAG|GTACCG and an acceptor sequence of GCAG|G. The sequence of 15 bases immediately preceding the acceptor, though not consecutive pyrimidine nucleotides, contains 60% pyrimidines. Some differences in the nucleotide sequences surrounding the GU and AG sites are not usual, as cryptic splice signals commonly show variation from consensus; this contributes to the signals being weaker than true splice signals (Roca et al., 2003).

The modTAT•Naglu cDNA construct, which no longer retained the strong endogenous splice signals of its progenitor pre-mRNA, likely underwent cryptic splicing when expressed in *Sf9* cells. RT-PCR using total RNA from *Sf9* cultures transfected with a mutagenized Naglu construct revealed that the alteration in three nucleotides, two at the cryptic splice donor site and one at the acceptor site, eliminated formation of the 426 bp splice product. Comparison of protein expression between cultures with the original and mutagenized modTAT•Naglu constructs is currently underway and will determine whether or not splicing had a substantial impact on expression levels.

Conclusion

RT-PCR on Naglu RNA harvested from transfected *Sf9* cells revealed a 426 bp non-target PCR product. This amplicon was shown to contain sequence corresponding to the 5' and 3' ends of Naglu mRNA, with a 1735 bp fragment missing in between. The product was thought to result from activation of a cryptic splice site in the Naglu cDNA construct. This hypothesis was supported by the putative donor and acceptor splice signals strongly conforming to consensus splice signals for GU-AG splicing, while additional experimentation using *C.therm.* polymerase indicated that the putative splice product was likely not an artifact of the RT-PCR procedure. Site-specific mutagenesis was successfully used to eliminate the observed cryptic splicing in *Sf9* cultures. While prevention of Naglu mRNA degradation through splicing will hopefully increase yields of the active enzyme, thus eliminating a major obstacle to Naglu expression for enzyme replacement therapy, the results of this study also provide further evidence for the occurrence of unfavorable activation of cryptic splice sites when using cDNA constructs in recombinant systems. In such expression systems where high yields are critical, "spurious" RT-PCR products should be analyzed as potential products of cryptic splicing.

ACKNOWLEDGMENT

This work was supported by the Natural Sciences and Engineering Research Council of Canada Discovery Grant #138216-09 to F.Y.M.C. and a Pre-Doctoral Summer Studentship Grant from the Canadian Society for Mucopolysaccharide and Related Diseases to S.N.T.

REFERENCES

Bandsmer, J.C. (2004). *Expression of α-N-acetylglucosaminidase fused to the HIV-1 protein transduction domain and a modified protein transduction domain* [master's thesis]. Victoria, BC: University of Victoria.

Bandsmer, J.C., Campbell, T.N., Cheyne, I. and Choy, F.Y.M. (2006). Expression of active α-N-acetylglucosaminidase/TAT chimerae in cultured Spodoptera frugiperda cells. *Protein Pept. Lett.*, 13, 353-356.

Bowman, R.R., Hu, W.S. and Pathak, V.K. (1998). Relative rates of retroviral reverse transcriptase template switching during RNA- and DNA-dependent DNA synthesis. *J. Virol.*, 72, 5198-5206.

Bukovac, S.W., Bagshaw, R.D., Rigat, B.A., Callahan, J.W., Clarke, J.T. and Mahuran, D.J. (2008). Cryptic splice site in the complementary DNA of glucocerebrosidase causes inefficient expression. *Anal. Biochem.*, 381, 276-278.

Burset, M., Seledtsov, I.A. and Solovyev, V.V. (2000). Analysis of canonical and non-canonical splice sites in mammalian genomes. *Nucleic Acids Res.*, 28, 4364-4375.

Chan, C.Y., Carmack, C.S., Long, D.D., Maliyekkel, A., Shao, Y., Roninson, I.B. and Ding, Y. (2009). A structural interpretation of the effect of GC-content on efficiency of RNA interference. *BMC Bioinformatics*, 10 (Suppl 1), S33.

Delviks, K.A. and Pathak, V.K. (1999). Effect of distance between homologous sequences and 3' homology on the frequency of retroviral reverse transcriptase template switching. *J. Virol.*, 73, 7923-7932.

Henke, W., Herdel, K., Jung, K., Schnorr, D. and Loening, S.A. (1997). Betaine improves the PCR amplification of GC-rich DNA sequences. *Nucleic Acids Res.*, 25, 3957-3958.

Kralovicova, J., Christensen, M.B. and Vorechovsky, I. (2005). Biased exon/intron distribution of cryptic and de novo 3' splice sites. *Nucleic Acids Res.*, 33, 4882-4898.

Mader, R.M., Schmidt, W.M., Sedivy, R., Rizovski, B., Braun, J., Kalipciyan, M., Exner, M., Steger, G.G. and Mueller, M.W. (2001). Reverse transcriptase template switching during reverse transcriptase-polymerase chain reaction: artificial generation of deletions in ribonucleotide reductase mRNA. *J. Lab. Clin. Med.*, 137, 422-428.

Mount, S.M. (1982). A catalogue of splice junction sequences. *Nucleic Acids Res.*, 10, 459-472.

Nelson, K.K. and Green, M.R. (1990). Mechanism for cryptic splice site activation during pre-mRNA splicing. *Proc. Natl. Acad. Sci. USA*, 87, 6253-6257.

Pfeifer, T.A., Guarna, M.M., Kwan, E.M., Lesnicki, G., Theilmann, D.A., Grigliatti, T.A. and Kilburn, D.G. (2001). Expression analysis of a modified factor X in stably transformed insect cell lines. *Protein Expr. Purif.*, 23, 233-241.

Roca, X., Sachidanandam, R. and Krainer, A.R. (2003). Intrinsic differences between authentic and cryptic 5' splice sites. *Nucleic Acids Res.*, 31, 6321-6333.

Schmidtchen, A., Greenberg, D., Zhao, H.G., Li, H.H., Huang, Y., Tieu, P., Zhao, H.Z., Cheng, S., Zhao, Z., Whitley, C.B., Di Natale, P. and Neufeld, E.F. (1998). NAGLU mutations underlying Sanfilippo syndrome type B. *Am. J. Hum. Genet.*, 62, 64-69.

Schwarze, S.R., Ho, A., Vocero-Akbani, A. and Dowdy, S.F. (1999). In vivo protein transduction: delivery of biologically active protein into the mouse. *Science,* 285, 1569-1572.

Sinclair, G., Pfeifer, T.A., Grigliatti, T.A. and Choy, F.Y. (2006). Secretion of human glucocerebrosidase from stable transformed insect cells using native signal sequences. *Biochem. Cell Biol.*, 84, 148-156.

van de Kamp, J.J., Niermeijer, M.F., von Figura, K. and Giesberts, M.A. (1981). Genetic heterogeneity and clinical variability in the Sanfilippo syndrome (types A, B, and C). *Clin. Genet.*, 20, 152-160.

Winship, P.R. (1989). An improved method for directly sequencing PCR amplified material using dimethyl sulphoxide. *Nucleic Acids Res.*, 17, 1266.

Zhang, J. and Temin, H.M. (1994). Retrovirus recombination depends on the length of sequence identity and is not error prone. *J. Virol.,* 68, 2409-2414.

Zhao, H.G., Aronovich, E.L. and Whitley, C.B. (1998). Genotype-phenotype correspondence in Sanfilippo syndrome type B. *Am. J. Hum. Genet.*, 62, 53-63.

REVIEW

In: Biochemistry and Molecular Biology Coming of Age ISBN: 978-1-63321-815-4
Editor: Mary Miu Yee Waye

Chapter 9

COMPARISONS OF EEG/QEEG NORMATIVE DATABASES AND REVIEW OF STUDIES OF NEUROFEEDBACK FOR DYSLEXIA

Warwick C. L. Cheung
Division of Biomedical Engineering,
The Chinese University of Hong Kong,
Shatin, N.T., Hong Kong SAR, China

ABSTRACT

This chapter analyzes and discusses the process of neurofeedback, particularly the considerations in selecting a normative EEG/QEEG database when it is used for people with dyslexia. Those considerations are partitioned into three main areas: technical considerations, analysis tools availability and statistic information. The details of those considerations are presented and discussed. Four major normative EEG/QEEG databases: BRID, HBI, NeuroGuide and SKIL, are reviewed and a side-by-side comparison is constructed. The side-by-side comparison is based on their technical specifications, analysis tools and statistic information. Another purpose of this chapter is to review the current status of neurofeedback for dyslexia. Eight studies of neurofeedback for dyslexia are reviewed and summarized.

INTRODUCTION

Developmental dyslexia (hereafter referred to as dyslexia), excluding the one acquired after birth, is a specific and significant impairment in the development of reading skills that is not solely accounted for by mental age, visual acuity problems, or inadequate schooling [1] Dyslexia is a common and important problem in all industrial societies, with a rate of 5–10%, for which no consistently effective treatment is available [2]. Neurofeedback is an operant conditioning procedure, in which an individual learns to modify the electrophysiological aspects of his/her own brain [3]. Nowadays, there are some studies of the efficacy of neurofeedback treatment on dyslexics. Some of them showed quite positive results in this kind of treatment. Those details will be presented in the later section of this chapter.

This chapter has two main objectives. The first one is to analyze and discuss the process of neurofeedback, particularly the considerations in selecting a normative EEG/QEEG database when it is used for dyslexics. The other is to review the current status of neurofeedback for dyslexia.

The main contents of this chapter are divided into five sections. The first one briefly presents the background of dyslexia, including its definition, brain abnormalities, theories of dyslexia and dyslexia in Chinese. The second one describes neurofeedback, including its principles, overview of the process, and the significance of a normative EEG/QEEG database in neurofeedback treatment. The third lists the considerations in selecting a normative EEG/QEEG database and each of the parameters are discussed. The fourth contains the results of comparisons of four major normative EEG/QEEG databases.

The last one reviews several studies of neurofeedback for dyslexia, in which the design of an experiment, the inclusion criteria of subjects, the treatment protocols and the results are summarized.

BACKGROUND OF DYSLEXIA

Definition

At present, dyslexia can only be diagnosed clinically. According to [1]. (World Health Organization, ICD-10 manual), "developmental dyslexia" is coded as "F81.0: Specific Reading Disorder".

The main feature is a specific and significant impairment in the development of reading skills (including reading comprehension skill, reading word recognition, oral reading skill, and tasks requiring reading) that is not solely accounted for by mental age, visual acuity problems, or inadequate schooling. It excludes those impairments of reading skills due to acquired neurological disorders.

It is frequently associated with disorders in speech or language development, spelling difficulties, and emotional and behavioral disturbances. Similarly, according to [4]. (American Psychiatric Association, DSM-IV-TR), "dyslexia" is coded as "315.00: Reading Disorder".

This is characterized by, "reading achievement (i.e. reading accuracy, speed, or comprehension as measured by individually administered standardized tests) that falls substantially below that expected given the individual's chronological age, measured intelligence, and age-appropriate education. The disturbance in reading significantly interferes with academic achievement or with activities of daily living that require reading skills. If a sensory deficit is present, the reading difficulties are in excess of those usually associated with it. It excludes those having a neurological or other general medical condition or sensory deficit." In general, dyslexia is defined as "an unexpected, specific, and persistent failure to acquire efficient reading skills despite conventional instruction, adequate intelligence, and sociocultural opportunity" [5].

Brain Abnormalities

For those studies that investigated the differences of brains between typical reading persons and those with dyslexia can be categorized into three main areas: structure, function and electrophysiology.

Structural Abnormalities

Structural abnormalities can be investigated by using the following technologies: magnetic resonance imaging (MRI), voxel-based morphometry (VBM) and diffusion tensor imaging (DTI). VBM is a tool to study local tissue concentrations, usually gray matter, on the voxel-level, which is obtained by performing a series of image processing procedures, including segmentation of each tissue, warping each tissue segment to a common anatomical space, and smoothing on a T1-weighted magnetic resonance image [6]. As a VBM analysis is obtained through a series of image processing procedures, the interpretation of this kind of study must be made with caution, especially the process of how to obtain the final VBM image from a magnetic resonance image [6]. DTI provides specific information about the structural properties and directional orientation of white matter tracts by indirectly measuring the properties of the diffusion process of water molecules in the brain [7]. Similar to VBM, DTI is obtained by applying a mathematic transformation on the data within a voxel acquired from MRI with special procedures enabling variations in diffusion of water molecules to be visualized [7]. From the mathematic transformation, more information can be derived to quantify the properties of white matter, which includes fractional anisotropy (FA), radial and axial diffusivity, and fibertracking. According to the descriptions found in [7], FA is an index of the amount of anisotropy, which is generally interpreted as "a quantitative biomarker of white matter integrity". Radial and axial diffusivity indices are the diffusivity in the direction perpendicular to and in the direction parallel to the principal axis of diffusion respectively. Fibertracking is a color image which is obtained by combining an FA image with information on the principal diffusion direction and it is possible to visualize the orientation of white matter. However, as emphasized in [7], the fibers illustrated in fibertracking cannot be considered to represent individual axons in reality because the voxel size (the minimum analysis unit obtained from MRI) is much greater than the size of microscopic factors of an axon, such as fiber crossing, kissing, and Y-shape branching.

From the reviews done by Caylak [8] and Habib [9], possible structural abnormalities associated with dyslexia can be summarized as follows. The first abnormality is the absence of leftward asymmetry of the cortical regions [8,9]. The second one is the increase in size of the corpus callosum [8,9]. And the last one is the decrease in size of the cerebellum [8]. No convergent conclusions can be drawn from the results of those MRI studies that were reviewed and concluded by [8] and [9].

With the advanced technology of VBM, the study of [10] identified significant gray matter differences in the right cerebellar anterior lobe, and right and left pars triangularis in dyslexic children. It also demonstrated gray matter volume differences in the left and right lingual gyrus, left inferior parietal lobule, and cerebellum in dyslexic children. According to [11], significantly reduced gray matter volumes were found in both temporal lobes in dyslexic individuals with VBM. It also found reduced gray matter density in the middle and inferior temporal gyri and increased gray matter density in the precentral gyri bilaterally in dyslexics. From the study of [12], reduced gray matter volume was found only in the left parietal region of the dyslexic group relative to both age-matched and reading-matched control groups. In the study of [13], it was found that dyslexic children had decreased white matter volumes in the left hemispheric network comprising the motor cortex, dorsal premotor cortex, ventral premotor cortex, and planum polare on the superior temporal gyrus. In the recent meta-analysis of nine VBM studies of developmental dyslexia including both adults and children [6], it was found that people with dyslexia had reduced gray matter volume in both the right superior temporal gyrus and the left superior temporal sulcus. However, in another study [14], which included people with specific language impairment (difficulties in production and reception of language), non-impaired siblings of those people, and unrelated control participants with typical language ability, it was showed that the group with specific language impairment had more gray matter in the left inferior frontal cortex and less gray matter in the right caudate nucleus and superior temporal cortex bilaterally when it was compared to the control group. It also demonstrated that the non-impaired siblings had less gray matter in the caudate nucleus bilaterally than that in the control group.

In the review of eleven DTI studies [7], it was concluded that people with dyslexia, including children and adults, lower FA values were observed in their left temporoparietal and frontal cortical regions. Moreover, the meta-analysis of ten DTI studies performed in [7], from the fibertracking, it indicated that those regions with lower FA values coincided with the left arcuate fasciculus and corona radiata, with a few studies supporting the posterior part of the corpus callosum that was also included. However, in another DTI study [15], it was demonstrated that children with dyslexia had significantly higher FA value in the posterior of corpus callosum, whereas the mean diffusivity and radial diffusivity were reduced substantially.

Functional Abnormalities

Functional abnormalities can be studied by using the following technologies: regional cerebral blood flow (rCBF) mapping, positron emission tomography (PET) and functional magnetic resonance imaging (fMRI). Those technologies are usually applied to test hypotheses of cortical functions in both people with typical reading ability and those with dyslexia [8]. Those investigation techniques involve some stimuli to be applied to subjects or tasks to be performed by subjects. During stimulation or when performing tasks, a functional brain image is captured, which shows the regional metabolic activities, e.g. oxygen consumption (PET and fMRI) or blood flow (rCBF), inside the brain. The change of the level or extent of those regional metabolic activities is directly related to the degree of cognitive activities of the brain and the locations of the change of metabolic activities are related to certain cognitive functions. Generally, the following cortical functions are usually tested: auditory processing (phonological processing), visual processing (processing of graphemes), working memory, phoneme-grapheme association, and awareness (attention). The tasks or stimuli used can be categorized into lexical (specific language related) and non-lexical (not relating to any languages, usually symbols or audible tones) ones. Because of the involvement of stimuli or tasks, the results strongly depend on what stimuli or tasks are applied. In addition, most experiments were designed to test their own hypotheses by using different stimuli and tasks [8]. Therefore, a variety of results and findings were obtained [5,8,9].

Nevertheless, some studies are worth being mentioned in this chapter. There was a co-localization study of abnormal brain structure and function in people with specific language impairment [14], which has been mentioned in the "*Structural Abnormalities*" section in this chapter. In terms of structural abnormalities, it showed that the group with specific language impairment had more gray matter in the left inferior frontal cortex and less gray matter in the right caudate nucleus and superior temporal cortex bilaterally when it was compared to the control group. It was also demonstrated that the non-impaired siblings had less gray matter in the caudate nucleus bilaterally than that in the control group. In the same study [14], those subjects were also tested with the auditory responsive naming task and the corresponding fMRI analysis was performed. It was showed that the group with impairment had reduced activation in the left inferior frontal cortex, right putamen, and in the superior temporal cortex bilaterally when it was compared to the results from the control group. It was also revealed that the non-impaired siblings had greater activation in the right cerebellar lobule VI during the speech condition of the task and there were no significant differences in other contrasts of the task when it was compared to the results from the control group.

Raschle, et al. [16] investigated functional networks during phonological processing in 36 prereading children, in which 18 children had familial risk of developmental dyslexia (i.e. at least one first-degree relative with a clinical diagnosis of developmental dyslexia) and 18 children did not had such familial risk. The fMRI results revealed that the children with familial risk of developmental dyslexia had reduced activation in bilateral occipitotemporal and left temporoparietal brain regions. This study showed that the functional abnormalities in people with dyslexia could be identified before the beginning of acquiring literacy. In another study done by Kovelman, et al. [17], they included four groups of children. The first one was a group of children with typical reading ability. The second one was a group of children with clinically diagnosed dyslexia. The third one was a group of age- and verbal IQ-matched control who had typical reading ability for a direct comparison to dyslexic children. And the last one was a control group of kindergarteners with at least average language ability, no family history of reading difficulties, younger than but matched for verbal IQ and on raw scores for phonological awareness with the dyslexic children. Functional MRI was used to identify the neural correlates of phonological awareness using an auditory word-rhyming task. It was showed that both typically developing children and kindergarteners recruited left dorsolateral prefrontal cortex when making explicit phonological judgments while the children with dyslexia did not.

In the meta-analysis done by Richlan, et al. [18], 18 functional imaging studies were examined, including those using fMRI and PET, for predominance of a phonological left temporoparietal dysfunction in dyslexic children (in 9 studies) and predominance of a visual-orthographic left occipitotemporal dysfunction in dyslexic adults (in 9 studies). This meta-analysis concluded four main points. Firstly, both dyslexic children and adults had dysfunction in left ventral occipitotemporal region. Secondly, it was inconclusive on the possible predominance of the left temporoparietal dysfunction in dyslexic children. Thirdly, only dyslexic adults, but not in children, showed under-activation in superior temporal regions. Fourthly, dyslexic children showed under-activation in bilateral inferior parietal regions, but this abnormality was no longer present when foci identified by higher dyslexic task-negative activation were excluded.

Electrophysiological Abnormalities

The electrophysiological abnormalities can be investigated by using electro-encephalography (EEG) or magneto-encephalography (MEG). EEG is brain-related electrical potential recorded from the scalp while MEG is a non-invasive technique for detecting magnetic fields that are associated with brain activities [19]. Similar to investigating functional abnormalities, electrophysiological abnormalities usually are investigated while subjects are being stimulated or performing tasks. Consequently, the results obtained from EEG or MEG also depend on what stimuli and tasks are applied [20,21] and what hypotheses are to be tested. Same as examining functional abnormalities, the following cortical functional tests are usually used in electro-physiological analyses: auditory processing (phonological processing), visual processing (processing of graphemes), working memory, phoneme-grapheme association, and awareness (attention). And both lexical (specific language related) and non-lexical (not relating to any languages, usually symbols or audible tones) materials will be used as tasks or stimuli for subjects in electrophysiological analyses.

Most EEG studies on people with dyslexia used event-related potentials (ERPs) to characterize the electrophysiological abnormalities found in dyslexics. ERP is the local field potential recorded during a psychological task and it is averaged over several trials of the same category which is associated with particular stages of information processing in the brain [19]. ERPs are suitable for investigations of the levels of cognitive processing required for reading because they provide sensitive neurophysiological measures of the timing and cortical utilization of different stages of cognitive processing [9]. Most of the ERP studies in dyslexics usually focus on electrophysiological events at P300, N400 or MMN (mismatch negativity) [8]. P300 is elicited in oddball tasks where subjects under test attend to a train of frequently occurring stimuli and respond to an infrequent irregular stimulus (oddball), and this is related to awareness and evaluation of stimuli as well as memory updating [9]. N400 is a negative component which occurs approximately 400 ms after stimulus is applied and it is usually associated with visual and auditory sentence comprehension tasks [22]. MMN was firstly described in [23], which is a negative component found in frontocentral region and it peaks between 100 and 250 ms after stimulus onset. It can be generated in an oddball paradigm where any physically deviant auditory stimulus occurs randomly and infrequently in a series of homogeneous stimuli and it can be elicited independent of attention. Also, same as the case in functional abnormalities studies, a variety of results and findings were obtained according to different hypotheses, stimuli and tasks. Detailed summaries can be found in studies [8] and [9].

Although there are a variety of electrophysiological analyses with different stimuli and tasks which tested different kinds of hypotheses, only those experiments with special design are mentioned in this chapter. In the study of [24], basic auditory processing was assessed by recording ERPs when paired tones were presented in an oddball paradigm. There were three groups of children to be tested. One group consisted of children with dyslexia and familial risk of dyslexia. Another one, control group, consisted of children with typical reading ability but they did not have familial risk of dyslexia. The remaining group consisted of children with typical reading ability and also had familial background of dyslexia. It was found that the control group and the group of typical reading children with familial risk of dyslexia showed larger ERP responses at the P1 and P2 time windows in the left hemisphere compared to the right for both short and long within-pair intervals of tones, whereas children with dyslexia showed this pattern only for the tone pairs with long within-pair interval, and for the tone pairs with short within-pair interval, children with dyslexia showed equal amplitudes over both hemispheres.

Another study [25] examined three groups of children also. One group consisted of children with clinically diagnosed dyslexia. Another group consisted of the siblings of those dyslexic children but they did not suffer from dyslexia. And the final one consisted of control children without relationship with the other groups. It measured the MMN component during a passive oddball paradigm which presented the consonant-vowel stimuli of /da/ and /ba/ to the subjects. It was demonstrated that dyslexic children and their unaffected siblings showed attenuated late MMN component (300-700 ms) when it was compared to that from the control children.

Dhar, et al. [26] focused on dyslexic adults. They measured the inter-hemispheric coherence of the dominant frequency during a visuo-spatial attention task. It was found that the dyslexic group showed reduced, and more diffuse, inter-hemispheric coherence of alpha (8.5 Hz) activity in the central-parietal cortex while no group differences in inter-hemispheric coherence were found on frontal, temporal or central sites.

Conclusion

The phenomenon of reduced activity in the left hemisphere and increased activity in the right hemisphere in people with dyslexia is the most consistently found result in different studies which utilized fMRI, PET or ERP analyses with different stimuli applied to subjects. It seems to indicate that people with dyslexia have difficulties in activating some cortical regions while they are processing linguistic information. However, the exact cortical regions with activation problems cannot be concluded at this moment since different stimuli and tasks will activate different brain regions, and it is difficult to accurately map a particular person's brain to a standardized brain template in order to locate the exact location of the abnormally activated regions in the standardized brain model. There is a trend to investigate whether the people with familial risk of dyslexia demonstrate the similar brain activation patterns, no matter they have clinically diagnosed dyslexia or not, while they are processing some language related stimuli or tasks. It is worth mentioning that at present, longitudinal study about the changes in the brain, no matter in structure, function, or electrophysiology, in people with dyslexia from childhood to adulthood is rare. This kind of study should be useful for investigating whether there is a compensation mechanism due to neuroplasticity.

Theories of Dyslexia

Several theories were proposed to describe the causes of dyslexia. These theories came from those studies with the purpose to elucidate how the dyslexic brain functions or malfunctions [9]. Habib [9] emphasized that considerable caution was required when there is any attempt to develop any explanatory model of dyslexia from the results of those studies, since most of them only reflected statistical tendencies and never represented a systematic rule.

The details of the theories of dyslexia are not discussed in this chapter. But the basic concept of each one are presented. Most of them are excerpts from the study of [27].

Phonological Theory

Phonological theory is the most commonly found and mentioned in the studies of dyslexia.

As stated in [27], phonological theory postulates that dyslexics have a specific problem in the representation, storage and/or retrieval of speech sounds. It explains dyslexics' reading impairment by appealing to the fact that learning to read an alphabetic system requires learning the grapheme-phoneme correspondence. The support for this theory comes from the evidence that dyslexic individuals perform phonological awareness tasks particularly badly.

Rapid Auditory Processing Theory (Temporal Processing Theory)

As summarized in [27], rapid auditory processing theory specifies that the deficit lies in the perception of short or rapidly varying sounds [28,29].

The support for this theory comes from the evidence that dyslexics show poor performance of a number of auditory tasks which include frequency discrimination [30,31] and temporal order judgment [28, 32].

Visual Theory

According to [27], visual theory, is summarized as postulating the magnocellular pathway which is selectively disrupted in certain dyslexic individuals, leading to deficiencies in visual processing, and via the posterior parietal cortex, to abnormal binocular control and visuospatial attention [33,34].

Evidence for magnocellular dysfunction comes from several areas of studies. One is from anatomical studies showing abnormalities of the magnocellular layers of the lateral geniculate nucleus [35].

Another one is from psychophysical studies showing decreased sensitivity in the low spatial frequencies and high temporal frequencies in dyslexics [36,37]. The final one is from some brain imaging studies in dyslexics [38].

Cerebellar Theory

As stated in [27], cerebellar theory postulates that the cerebellum of a person with dyslexia is mildly dysfunctional and then a number of cognitive difficulties in speech articulation, phonological representation and the learning of grapheme-phoneme correspondences consequently appear [39,40].

The support for this theory comes from the evidence of poor performance of dyslexics in a large number of motor tasks [41], in dual tasks demonstrating impaired automatization of balance [39], and in time estimation which is a non-motor cerebellar task [42].

Magnocellular Theory

Magnocellular theory was concisely stated in [27]. as a theory that tries to postulate the magnocellular dysfunction which is not restricted to the visual pathways but also causes impairment, generally, in all modalities, including visual, auditory and tactile [33].

Double-Deficit Theory

As stated in [43], double-deficit theory proposes that reading disabilities can be the result of poor phonological awareness and/or automatic naming skills. Poor phonological awareness refers to disabilities in identifying and manipulating sounds in speech, whereas poor automatic naming implies the disability to translate graphemes to phonemes.

Conclusion

In reality, it is possible to find more than one impairment in a person with dyslexia. From the results made in [27], it concluded that although the phonological deficit could even arise independently of any sensory or motor impairment, a significant proportion of dyslexics suffered from other sensory or motor disorders also. From the analysis of [5], there is no single theory that can conclude all phenomena found in dyslexics. Caylak [8] concluded that dyslexia appears as a complex language disorder, which involves many brain areas and a variety of pathophysiological mechanisms. Therefore, it seems that it would be improper to conclude the cause of dyslexia solely by any one of the sensory or motor disorders. A more appropriate approach is to find out what impairments exist in people with dyslexia and then identify what impairments hinder the dyslexics from learning a language and how those impairments interfere with the language learning process.

DYSLEXIA IN CHINESE

As stated in [44], unlike alphabetic writing systems that follow a design principle of mapping graphemes onto phonemes, the Chinese logographic system maps graphic forms (characters) onto morphemes and the phonology of written Chinese is defined at the monosyllabic level, with no parts of a character corresponding to phonemes. This means that Chinese language is a morphosyllabic writing system [45,46]. Chinese features are much more arbitrarily mapped between characters and oral language than English features [47]. So the letter-sound conversion rules that exist in all alphabetic languages cannot be applied to Chinese language [44]. Research has shown that recurrent and parallel activations of three interconnected linguistic constituents: orthography, meaning and phonology are involved in Chinese reading [48,49]. Therefore, it is believed that the cortical activation irregularity associated with impaired Chinese reading is different from that associated with impaired English reading [44].

As Chinese language consists of visually distinct characters, visual processing may play an important role in learning to read Chinese [47]. Research in Hong Kong has shown that a fairly large subset of dyslexic Chinese readers exhibited orthographic and visual processing deficits [50,51]. According to the study in [47], which was done in Hong Kong, from hierarchical regression analyses, it was revealed that visual processing, but not auditory temporal processing, contributed unique variance to Chinese character recognition even with other cognitive measures controlled. From another study [44], by using fMRI, it was demonstrated that Chinese dyslexics had reduced activation of LMFG (left middle frontal gyrus) at Brodmann area 9 where LMFG carries out the representation and working memory of visuospatial and verbal information, and coordinates cognitive resources as a central executive system [52,53,54]. According to [55], it was showed that Chinese dyslexics exhibited reduced gray matter volume in the LMFG region by using VBM and reduced activation in the same region by using fMRI with a phonological processing task (character rhyme judgment). Another study from Hong Kong [56] examined group differences in processing of orthographic information between twelve dyslexic children and eleven children with typical reading ability by analyzing their N400 ERP responses. It was demonstrated that the typical reading children showed a more negative N400 component in the pseudo-character condition (the structure of semantic and phonetic radicals was in correct orthographic order) when it was compared to the case of non-character condition (the structure of the semantic and phonetic radicals was reversed). However, the dyslexic children did not show any differences in N400 component across those two experimental conditions.

In summary, the number of studies about dyslexia in Chinese people is limited. It is difficult to conclude that Chinese people with dyslexia exhibit different impairments from the people who use alphabetic writing systems. And, as stated before, it would be improper to conclude the cause of dyslexia solely by any one of the sensory or motor disorders. More investigations into what impairments hinder the dyslexics from learning a language and how those impairments block the language learning process should be done instead.

NEUROFEEDBACK

According to the explanation in [3], "neurofeedback is a sophisticated form of biofeedback based on specific aspects of cortical activity (e.g. EEG). It refers to an operant conditioning paradigm in which an individual learns to modify the electrophysiological aspects of his/her own brain and the goal is to teach the individual what specific states of cortical arousal feel like and how to activate such states voluntarily." This is a self-regulation skill that the trainee learns to regulate aspects of his/her autonomic nervous system functions [57]. In practice, neurofeedback is described in [57] as follows:

> "Neurofeedback training requires attaching sensors to the scalp for the purpose of acquiring electrophysiological signals produced by the brain. That signals are fed to the trainee with the goal of gaining mental control over subconscious biological processes.

The trainee receives moment-by-moment information about changes beneath the sensors. Information may come in the form of auditory tones or graphics display."

EEG signal is not the only one signal that can be used for monitoring the brain's activities. Hemoencephalography (HEG) signal that detects the cerebral blood flow can be used instead [57]. When the trainee's EEG signal meets the desired goal, the trainee is presented with a reward in the form of selected sounds and displays [58]. When the trainee's EEG signal produces a value that is not desired, a different sound or visual image is presented to the trainee to inhibit that particular signal [58]. Despite sound and visual image, tactile modality can also be used as a form of feedback to the trainee [19]. The exact physiological processes involved in changing the pattern of cortical activity are not well understood [3,58]. Nonetheless, research has shown that healthy individuals [59,60]., people suffering brain injury [61,62], epileptics [63,64,65] and schizophrenics [66] have been able to exhibit changes in their cortical activity following neurofeedback training.

Although the exact physiological processes involved in a neurofeedback training are not well understood [3,58], its success is strongly believed to be based on two facts. Firstly, the brain states are objectively reflected in parameters of EEG recorded from the scalp [19]. Secondly, the brain has plasticity (neuroplasticity) to memorize the desired state of the brain through a series of feedback training [19]. For the first point, it is supported by a lot of studies that showed the correlation between EEG features and different cognitive states. For frequency domain analysis, an EEG signal has been partitioned into several frequency bands that are associated with different cognitive states. For example, Delta band (0.32 Hz – 4 Hz) is associated with sleep; Theta band (4 Hz – 8 Hz) is associated with creative and spontaneous state; Alpha 1 band (8 Hz – 10 Hz) is a reflection of an unaware state of relaxation; Alpha 2 band (10 Hz – 13 Hz) is associated with a state of awake and alert; Low Beta band (13 Hz – 21 Hz) is a reflection of activation of the cortex; High Beta band (22 Hz – 35 Hz) is associated with higher levels of concentration [67]. For time domain analysis, event-related potentials (ERPs) are commonly used. As stated before, ERPs are evoked potentials measured on the scalp after a special stimulus has been applied to the subject [19]. They are analyzed in time domain and different peaks and valleys that appear at particular time corresponding to different cognitive processes [22]. For detailed descriptions of how EEG features correlate with different cognitive states, the following references are recommended: [19], [57], [67] and [22]. For the second point, neuroplasticity is a hypothesis that connections and effectiveness of chemical synapses inside the brain can be changed through experience and learning [68]. But neurofeedback can be effective only if the trainee is able to learn a new skill and willing to accept the feedback [57].

As stated in [57], a neurofeedback training plan or protocol requires information from at least three clinical areas: diagnosis, EEG data or brain map (pre-treatment) and an understanding of neurology. Diagnosis process involves a psychiatric interview and a psychosocial interview with a brief medical history. The EEG data measured before treatment provide a baseline of a client's brain states. These data are usually transformed to some parameters, e.g. frequency components, and then those parameters are compared with those in a normative database to find out what electrophysiological abnormalities exist in the client. Both the results from diagnosis and pre-treatment EEG data are analyzed and diagnosed by an experienced clinician. The clinician uses his/her knowledge and experience in neurology to diagnose the problem and then drafts a neurofeedback treatment plan for the client. Also, the treatment plan can be adjusted while treatment is being executed in response to the progress of the trainee.

For a neurofeedback treatment, there are three main components: an EEG capturing and processing unit, a decision process, and a result feedback path as shown in Figure 1. The EEG capturing and processing unit consists of sensors, amplifiers, filters, analog-to-digital converters and other necessary signal processing units, e.g. noise suppression, artifact removal, Fast Fourier Transform (FFT), etc.

Those captured EEG data are usually transformed to some quantitative parameters or features, e.g. frequency components by using FFT, and then those quantitative parameters or features are passed to the decision process. Inside the decision process, there is a normative EEG/QEEG database which provides a set of normal references for comparing the incoming features. By using some statistical methods, the incoming features will be classified as "normal" or "abnormal" case.

If they are classified as a "normal" case, a "reward" will be sent to the trainee to encourage the trainee in order to reinforce this "normal" brain state. Otherwise, an "inhibition" will be sent to the trainee to discourage the trainee in order to inhibit the activation of this "abnormal" brain state. The "reward" or "inhibition" status will be transformed to one of the following forms: visual, auditory or tactile form that can be sensed by the trainee. The sensation of "reward" and "inhibition" should be distinct, impressive and not boring [57]. However, the sensation of "reward" and "inhibition" is very subjective. It totally depends on the trainees.

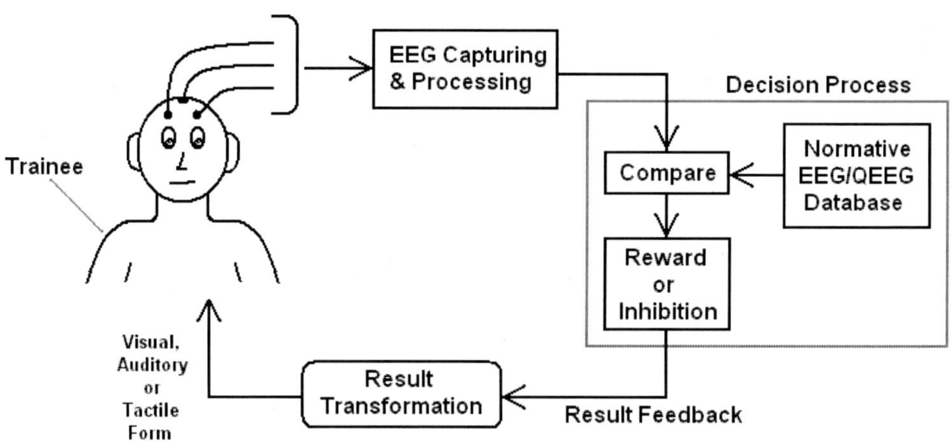

Figure 1. Basic setup of the process of neurofeedback.

Neurofeedback treatment can be executed at rest or with active tasks intervention. For treatment at rest, the trainee sits under a comfortable environment. His/her EEG signals are measured either in eye-closed or eye-open condition [57]. This technique is the most fundamental neurofeedback process. With active tasks intervention, some stimuli or active tasks are applied to the trainee. EEG signals are captured while the trainee is being stimulated or is performing a special task.

From the basic setup of the process of neurofeedback, the normative EEG/QEEG database plays an important role in the process. It directly controls the determination of "normal" and "abnormal" brain states. This is the heart of neurofeedback process. In practice, two functions are provided by a normative EEG/QEEG database. On one hand, it provides a "normal" reference for classifying brain states [19,57]. On the other hand, it provides analysis tools which are used to extract some quantitative features from the trainee's EEG signals and then to compare with those features stored inside the database in order to make a decision of "normal" or "abnormal" during neurofeedback process. Commercial database providers usually include a set of analysis tools that are specifically designed for their own databases. For example, if a database includes a set of "normal" references when the subject is performing an active task, e.g. recognizing a set of words, the provider must include the same active task (or paradigm) together with the database for clinicians. Some database providers even include an inverse analysis that is used to find out the location of electrical activities occurred inside the brain. The criteria in selecting a normative database will be discussed in detail in the next section.

CONSIDERATIONS IN SELECTING A NORMATIVE EEG/QEEG DATABASE

As stated in the previous section, a normative EEG/QEEG database is necessary to provide a "normal" reference [19,57] and to provide analysis tools. In order to obtain a meaningful comparison with a normative database, the conditions used in practice must be as similar as the conditions used in generating the normative database. The conditions include the age and ethnics of recorded subjects, the EEG measuring equipment setup, signal processing applied to the captured EEG signals, removal of artifacts, under eye-closed, eye-open or tasks performing, etc. So it is necessary for a normative database provider to disclose the procedures of generating the database as detailed as possible. Also, as mentioned in the previous section, some analysis tools are provided by the database providers that are specifically designed for their own database. There is a strong relationship between the database data and the provided analysis tools. Therefore, when a clinician uses a particular database, that particular database implicitly restricts the following: EEG capturing methods and equipments, the analysis tools and methods, and ethnics and age of subjects. Ideally, a normative database should represent the range of "normal" within the whole population of individuals which are rigorously screened in order to include those "healthy" or "normal" individuals only [19]. However, as stated by Kropotov [19], there is no database that can fit this condition practically. Consequently, the currently available databases are more appropriately considered as "reference" instead of "normative" databases that provide significant information for clinicians to assess the patient's abnormalities [19]. Because no ideal normative database can fit for all situations, the selection of using which normative database becomes a necessary process in order to fit the requirements of each particular case.

In this section, the considerations in selecting a normative EEG/QEEG database have been classified into 3 main parts: technical considerations, analysis tools availability, and statistic information. Technical considerations are those specifications that are related to how EEG signals are captured and how they are modified and processed. Analysis tools availability reflects what EEG/QEEG quantitative features are stored inside the database and what their corresponding analysis tools are included. Statistic information indicate what subjects are included in the database, how they are screened, the sensitivity, specificity and reliability of using a parametric Z-score method to differentiate "normal" and "abnormal" signals and how to handle the time-of-day factors.

TECHNICAL CONSIDERATIONS

Number of Electrodes

In most clinical applications, 19 recording electrodes according to the International 10-20 system are used [57]. However, it must be noticed that placing electrodes on the scalp to capture electrical signals is a kind of sampling in space. This means that spatial aliasing will occur if the density of electrodes does not satisfy the Nyquist criterion that requires at least two spatial samples per minimum spatial wavelength [69]. But, at this moment, the optimum number of electrodes remains somewhat of an open question that deserves further attention because we do not know what the highest spatial frequency of electrical signal will appear on the scalp [69]. On the other hand, there are some emerging technologies that can approximately locate the sources of electrical activities by using electrical signals measured from the scalp, e.g. LORETA. The spatial resolution of those approximated sources depends on the number of electrodes to be placed on the scalp and more electrodes will give a finer tomography [70]. In addition, for some special analysis methods, e.g. common average reference and surface Laplacian, more number of electrodes are required [19,69].

Montages

As stated by Kropotov [19], in EEG machines with differential amplifiers, electrode potentials being measured are in reference to some other potential or potentials. There are many schemes of EEG measurements by using different references and all of them are commonly called montages. The varieties of montage introduce the following problem described by Kropotov [19]:

> "Ideally, the reference electrode must be electrically neutral, that is, its potential does not change throughout the recording. Unfortunately, due to the conductivity of the scalp, all locations on the head are in a sense electrically active and there is no place in which a reference electrode could be attached to meet the criteria of the ideal reference. This means that the electrical potential recorded from any single electrode will depend on the reference."

This means that EEG signals recorded with different montages will have different results. Therefore, when a particular normative EEG/QEEG database has been used in a neurofeedback treatment, the clinician must use the same montage as that used by the selected database to capture the trainee's EEG signal in order to have a meaningful comparison with the database. Several montages or referencing schemes will be described and discussed next.

The simplest one is called "bipolar" recording or "sequential" recording. The term "bipolar recording" often refers to the special case where the potential difference between two closely spaced electrodes is measured [69]. As the electrodes of any pair are moved closer together, they provide progressively better estimates of the local gradient of the potential along the scalp surface in the direction between the electrodes and then it can be an effective strategy to identify local generators [69].

"Linked-ears" montage is measuring electrodes' potentials in reference to two linked electrodes located at the left and right earlobes and similar approach can be applied to "linked-mastoids" montage also [19]. Most normative databases used linked-ears montage to capture EEG signals.

There are two ways to implement the linked-ears or linked-mastoids reference [69]. One approach is to physically link the ears (or mastoids) by a wire to serve as the reference. Alternatively, all scalp potentials are recorded with respect to a reference placed at one ear ($V - V_{E1}$), with the second ear potential measured with respect to the first ear ($V_{E2} - V_{E1}$). The mathematically linked-ears reference is calculated by subtracting half this potential difference from all the measured potentials, creating an artificial reference at the average potential of the two ears, that is:

$$(V - V_{E1}) - \frac{(V_{E2} - V_{E1})}{2} \equiv V - \frac{V_{E1} + V_{E2}}{2}$$

There is a potential problem of using physically linked-ears reference. As stated in [69], if electrode impedances of those two ears (electrodes) change, the reference may become a random reference, switching back and forth from ear to ear. This problem must be aware in practice.

"Common average" or "average" montage is a mathematical procedure that all electrodes' potentials are averaged and that averaged potential is used as the reference for all electrodes [19,69]. This common average montage would be the perfect reference if the head was a sphere and electrodes could be placed on the whole surface of this sphere [19]. However, in reality, these two conditions cannot be achieved. Moreover, as stated in [69], the effectiveness of the common average depends on the number of electrodes used to record the potentials, and the nature and number of sources. It is effective only if there are a sufficient number of electrodes (64 to 128 or more) distributed over the entire scalp [69]. This technique also has distortions when displaying generalized changes, especially generalized paroxysmal activity [71]. Therefore, this montage must be used with great care.

"Local average" montage is another mathematical procedure, in which a small number of electrodes in the vicinity of a target electrode are used for computing the mathematical reference [19]. As suggested in [19], the potential of the i-th electrode can be calculated by:

$$V_i' = V_i - \sum w_{ij} \times V_j$$

where, V_i is the measured potential at the i-th electrode, V_i' is the potential of the i-th electrode after converting to local average montage, and w_{ij} are the weights and they characterize the calculated reference. For example, the common average montage can be calculated by setting w_{ij} equal to $1/total\ no.of\ electrodes$. Other examples of local average montage are Laplacian montage and Hjorth montage. The Laplacian montage is best suited for localization of relatively focal sources but it has some limitations [19]. Firstly, its accuracy is highly dependent on the number of electrodes and the distance between electrodes. Secondly, the potentials at the periphery of the measurement area cannot be reliably estimated since electrodes at these points are not completely surrounded by an array of electrodes. Thirdly, it will significantly distort the location of diffuse sources.

In conclusion, there is no single montage that outperforms the others. For analysis, the choice of montage depends on what analysis will be done. In the study of [72], it evaluated how the linked-ear reference, single reference (bipolar), average reference and Laplacian reference affected the coherence and phase difference analyses under different signal-to-noise conditions by simulation. It was demonstrated that linked-ears and single references provided valid and systematic relations between the magnitude of coherence and signal-to-noise ratio. In contrast, the average reference and Laplacian reference failed to systematically vary as a function of the signal-to-noise ratio. On the other hand, if there are enough electrodes, common average will approach to the ideal reference condition [19,69]. For the neurofeedback treatment, the choice of montage has been selected by the database provider that users must follow the same montage.

Specifications of EEG Measuring Equipments

For the specifications of EEG measuring equipments used in generating the normative EEG/QEEG database, there are at least 6 parameters that are critical: electrode impedance, amplifier matching, sampling rate, bit resolution, bandwidth and signal-to-noise ratio. These are the specifications of the components from electrodes to the point after analog-to-digital conversion. The setup of this front-end affects the quality of EEG signals received.

Electrode impedance specified in the database must be ensured during practice. Otherwise, the amplitude of the received signals will be different. It affects those comparisons that use absolute values of received EEG signals. In order to equilibrate the EEG machine used in generating database and the one used in practice, an amplifier matching procedure is required [20]. Amplifier matching is a procedure to match the frequency responses of the EEG machine used in generating database and the one used in practice [20]. After knowing the difference between two systems' responses, scaling coefficients will be applied to all received EEG signals when it is used in practice in order to mimic the response of the EEG machine used in generating database data [20]. The bandwidth of the EEG machine used in generating database must be known because this limits the frequency range of EEG that can be analyzed. The sampling rate, bit resolution and signal-to-noise ratio of the EEG machine used in generating database determine the quality of data stored in the database.

Artifact Removal

EEG signal can be contaminated by various artifacts, such as eye movement artifact, muscle movement artifact, ECG artifact, cardio-ballistic artifact, and other non-brain sources of electrical activity [19]. For the case that the EEG signals are obtained under active tasks intervention, the problem of artifacts becomes more critical and serious. The most basic artifact removal technique is by using visual inspection that is done by an experienced EEG technician [20,71,73]. The experienced technician will eliminate those EEG signals that are contaminated by artifacts. One automatic method to remove artifacts of eye movement, muscle movement and ECG is to record concurrently those signals and then by using adaptive filtering approach to subtract those artifacts from the EEG signals [19,71]. Despite using adaptive filtering approach, Kropotov [19] suggested to use the ICA (Independent Component Analysis) [74] approach. As described in [19], the basic idea of ICA approach is the decomposition of the EEG signal into two components: the one that corresponds to neuronal electric activity; the other one that corresponds to artifacts. Each component consists of a waveform, describing the time course of the modeled activity, and a topography vector, describing how the waveform contributes to each recorded signal. Artifact activity can be reconstructed as a product of artifact topographies and waveforms. Of course, automatic artifacts removal techniques are useful especially under a neurofeedback treatment with active tasks intervention. However, the authors of [20] stated that those artifact removal techniques using ICA and PCA (Principal Component Analysis) approaches should be avoided because they will invalidate the computation of coherence and phase differences, which is a consequence of the regression and reconstruction of those raw EEG samples. It is necessary to pay attention on the integrity of the EEG signals and what analysis will be done after artifact removal.

Analysis Tools Availability

Eye-Closed/Eye-Open

This refers to the status of the subject while his/her EEG signals are being measured. For both "eye-closed" and "eye-open" conditions, EEG signals are captured when the subjects are at rest with states of closed eyes and open eyes respectively. Neurofeedback treatment under eye-closed or eye-open condition is the most basic treatment. The normative EEG/QEEG database used in neurofeedback must have data for both conditions.

Active Tasks

Authors of [20] stated the following about active tasks:

> "An active task refers to the recording of EEG and/or evoked potentials (EPs) while a subject performs some kind of perceptual or cognitive task. Many EEG, EP and ERP studies have reported reproducible changes in brain dynamics which are task dependent. Such studies are important for understanding normal and pathological brain processes responsible for perceptual and cognitive function. Active tasks are dependent on the intensity of stimuli, the background noise of the room, the distance between the subject and the stimuli, the subject's understanding of the task instructions, the subject's motivation, etc. These are very difficult to control across experimenters or across clinics for the purposes of constructing a reference normative EEG database."

From the above statement, two points can be noticed. Firstly, EEG studies with active tasks are important for understanding normal and pathological brain processes responsible for perceptual and cognitive function [20].

This means that it is also useful for understanding the problems in dyslexics who have specific problems while reading. Together with real-time LORETA analysis, that will be discussed later, the cortical abnormal activities existed in dyslexics while reading can be found or estimated.

With such kind of difference, a "reward" and "inhibition" scheme can be established for dyslexics during a neurofeedback treatment. From the comparison study of the effectiveness of neurofeedback treatment between standard (at rest) and active tasks intervention in treating reading disability and traumatic brain injury (TBI), it showed that the improvement of reading and auditory memory by using neurofeedback treatment with active task intervention was greater than that by using standard (at rest) neurofeedback treatment [58]. From the study in [21], it was concluded that a cognitive activation database is clinically useful.

Secondly, EEG recording under active tasks is dependent on the nature of stimuli and the subject's understanding of the task instruction [20]. It was also agreed by the authors in the study of [21]. This means that if a database includes a set of normative EEG data that are recorded under a particular activation task or paradigm, that active task or paradigm must be replicated as exact as the original one when it is applied to the neurofeedback treatment in practice. Otherwise, the decision process for classifying "normal" and "abnormal" cases will not be accurate.

Because of stimuli dependent, the design of active tasks or paradigms becomes important and critical. The efficacy of different paradigms used in neurofeedback treatment needs to be investigated.

At this moment, such kind of study is rare. Moreover, for those active tasks involving languages, e.g. auditory and visual words recognitions, this makes the normative EEG/QEEG databases become language or cultural specific. It may not be possible for one normative database to include all languages in the world. This problem increases the difficulty in generating a cognitive activation database.

Time Series Analysis

The fundamental EEG analysis is the brain waves analysis (i.e. analysis of alpha, beta, theta, etc.), which is obtained by passing the artifact removed EEG signal to several band-pass filters. Because of the different definitions of bandwidths of different brain waves, it is better to allow a user to define the bandwidth for analysis. A time series analysis focuses on a particular characteristic of the captured signal, such as amplitude, phase or particular range of frequency (alpha, beta, theta, etc.), which occurs in a particular region of the brain (scalp) on particular time. Event-related potentials (ERPs), event-related desynchronization (ERD) and event-related synchronization (ERS) are used for analyzing cognitive functioning after being stimulated. ERPs are local field potentials recorded during a psychological task and averaged over trials of the same category that are associated with stages of information processing in the brain [19]. ERD or ERS is a parameter that measures the percentage of decrease or increase (respectively) in the EEG power in a given frequency band within a given time interval in response to a given event [19]. These techniques are used for studying EEG signals in time domain under certain active tasks.

Spectral Analysis

It is a kind of frequency domain analysis of EEG signal by applying a series of mathematical processing. The process will be done after the EEG signal has been digitized and it usually involves the following steps:

1. Determine the duration of a sliding window.
2. Apply the sliding window on the EEG signal (i.e. extract a portion of EEG signal, in which the number of elements is the same as that of the defined sliding window) and this segment of signal is called an epoch.
3. Multiply the segment of signal by a window function. A window function is a numerical series containing the same number of elements as the number of signal samples in the epoch and the element values tend toward zero at both ends and toward unity in the middle [19].
4. Apply Fast Fourier Transform (FFT) on the result of Step 3 to obtain the frequency spectrum of the current epoch.
5. Move the sliding window to the successive EEG data. Usually, the sliding window will be moved 25% to 50% of its length. This will create 75% to 50% overlap of successive segments of data.
6. Repeat from the Step 2 until the end of EEG data.

From the steps listed above, there are three main factors that affect the quality of the spectral analysis results. The first one is the duration of the sliding window used in Step 1 (i.e. the number of elements in one epoch). The second one is the choice of window function used in Step 3.

And the third one is the number of points of FFT calculated in Step 4. The number of FFT points determines the theoretical frequency resolution of the results while the duration of the sliding window and the choice of window function determine the effective frequency resolution of the results. Usually the specification will specify the theoretical one, for example, 1 Hz (single bin) or 0.5 Hz. But it is seldom to mention what window function has been used.

Different window functions have different extent of spectral leakage that will introduce spurious peaks or conceal true peaks in the spectrum [75]. At the same time, applying a window function as indicated in Step 3 also introduces a widened main lobe of each frequency point in the spectrum that is calculated by FFT. This will cause the energy of each individual frequency point to spread into its adjacent frequencies, and this phenomenon is called spectral smearing [75].

This spectral smearing results in the decrease of the effective frequency resolution of the spectral analysis. The extent of the spreading of energy due to the widened main lobe can be reduced by increasing the duration of the sliding window used in Step 1. This means that the effective frequency resolution can be increased by increasing the length of analysis window. However, increasing the duration of the sliding window means to decrease the time resolution of the analysis, which may result in masking some short-term EEG signals. This time-frequency resolution tradeoff is the characteristic of this kind of spectral analysis method. In addition, the percentage of overlapping two adjacent sliding windows affects the smoothness of the resulting spectrum change with time.

Measurements – Absolute vs. Relative

Absolute values of amplitude or power refer to the data that come from the measurement directly while the relative one is a percentage measure. For example, relative amplitude may be defined as amplitude in a frequency band divided by total amplitude (i.e. the sum of amplitude in all frequencies) [73]. Absolute amplitude and power measures not only reflect the amplitude of brain-generated EEG, but also reflect non-brain factors such as scalp resistance, skull thickness and various anisotropic conductance properties of the skull, dura and scalp [76,77].

A typical method to control for differences in scalp resistance and skull thickness, etc. is to calculate relative amplitude and/or power [73]. However, in clinical practice especially in pre-treatment analysis, both absolute and relative brain maps will be considered but the judgments will not be solely based on the absolute reading [57].

Difference Measurement

It is the calculation of the difference of amplitude or power between two arbitrary electrodes.

Ratio Measurement

It is the calculation of the ratio of two arbitrary frequency bands, e.g. alpha/beta, theta/alpha, theta/beta, etc. This calculation is useful for some neurofeedback treatments that target the change of alpha/theta ratio [57], theta/beta ratio [78], theta/alpha [79,80], etc.

Coherence and Phase Analysis

EEG coherence is a measure of the degree of association or coupling of frequency spectra between two different time series while EEG phase delay is a measure of the temporal lead or lag of spectra, and both can be applied to studies of cognition [72]. As suggested by Schummer [81], coherence-training neurofeedback is a powerful tool that can teach the brain to facilitate or inhibit interneuron connectivity. Mathematically, coherence is defined as the normalized cross-power spectrum and phase delay as the phase angle between two simultaneously recorded EEG signals from different scalp locations per frequency band [72]. However, there is a lack of the agreement to the definition of calculation of coherence and phase delay [81]. Different hardware and software companies have employed different algorithms in calculating these values [21]. So it is necessary to make sure how coherence and phase delay are calculated before using them for analysis and neurofeedback treatment.

Inverse Calculation

A fundamental element of an EEG generator can be mathematically modeled as a dipole located within the cortex and it is oriented perpendicular to the surface of the cortex [19]. The solution to the inverse problem is to find out the position, orientation, and strength of the dipole that can fit the measured EEG signals [19]. The theoretical solution of the inverse problem is not unique, which means that there is infinite number of different source configurations that can fit the measured electric field [19,69,70,82,83]. Because of the non-uniqueness of the inverse problem, any inverse solution must depend partly on additional information, which includes solution constraints [69,82]. Although, there are some inverse tomographic representations available, all of them are not precise solutions and they are just approximations of the reality [19]. It is recommended to consult the study of [82] which reviewed several inverse solutions.

Nonetheless, Kropotov [19] has suggested an approach that can give better accuracy in localizing test sources which is called LORETA (Low Resolution Brain Electromagnetic Tomography). From the review in [82], it was concluded that in most cases, the LORETA solution gave satisfactory results in localizing a source. It was also demonstrated that for single source localization under different noise levels and different simulated source depths, the regularized sLORETA (Standardized Low Resolution Brain Electromagnetic Tomography) gave the best solution in terms of both localization error and ghost sources. The study of [84] demonstrated that sLORETA had exact, zero error localization when reconstructing single sources in all noise free simulations. A brief description of LORETA will be presented here. For detailed descriptions and its derivation can be found in [70], [83], [84] and the following URL: http://www.uzh.ch/keyinst/loreta.htm

LORETA was firstly introduced in [70] for localizing the electric activity (in terms of current density) in the brain based on multichannel surface EEG recordings. According to [70], there are three assumptions when using LORETA. Firstly, the neighboring neurons are simultaneously and synchronously activated. Secondly, the computational task is to select the smoothest of all possible 3-dimensional current distributions. Thirdly, a 3-shell spherical head model with a radius of 78 mm and equal inter-electrode distances on the scalp is assumed.

There are two main advantages of using LORETA, sLORETA and eLORETA over other inverse solutions. Firstly, they do not assume a limited number of dipolar point sources nor a distribution on a given known surface [70]. Secondly, they do not depend on the reference electrode [83]. Moreover, parametric statistics that rely upon a Gaussian distribution have been successfully used in studies of LORETA [85,86,87,88,89,90,91]. And from the study of [86], it was demonstrated that Z-score statistical analyses, using a LORETA distributed source solution, provided adequate sensitivity and validity. These results support that the results from LORETA analysis can be used to build a normative EEG/QEEG database with adequate sensitivity and validity. Once such database has been built, it can be applied to a neurofeedback treatment. As stated earlier, active tasks intervention with real-time LORETA analysis can provide the information of electrophysiological abnormalities occurred in the dyslexics while reading. With such kind of difference, a "reward" and "inhibition" scheme can be established for dyslexics during a neurofeedback treatment. Therefore, it is believed that dyslexics can be greatly benefited from this kind of neurofeedback treatment. Some studies [92,93,94] have demonstrated the possibility of using a LORETA normative database to execute neurofeedback treatments.

Maximum Processing Latency

It is the maximum time required to process EEG/QEEG data before getting results. For analysis, it usually refers to the processing latency along the path from sensors to the result after comparing with data stored in a normative EEG/QEEG database as indicated in Figure 2. For a neurofeedback treatment, it should refer to the processing latency along the whole path: from sensors to the result sensed by the trainee as indicated in Figure 3.

Obviously, this latency is both hardware and software dependent. This means that the latency depends on what processing algorithms and what machines are used. Therefore, it will be good if there is a testing procedure for obtaining this parameter when a new neurofeedback system is being deployed.

This procedure is necessary because of the nature of neurofeedback treatment which returns either "reward" or "inhibition" according to the "current" brain state. If the result is fed back to the trainee after his/her brain state has changed, this will incorrectly "reward" an abnormal brain state or "inhibit" a normal brain state. This kind of incorrect information violates the objective of neurofeedback. However, there is no rule or guidelines to specify what should be the maximum.

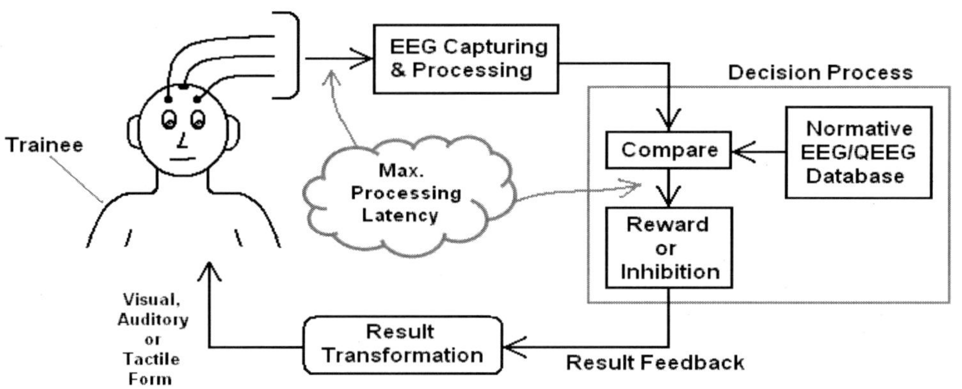

Figure 2. Maximum processing latency of analysis.

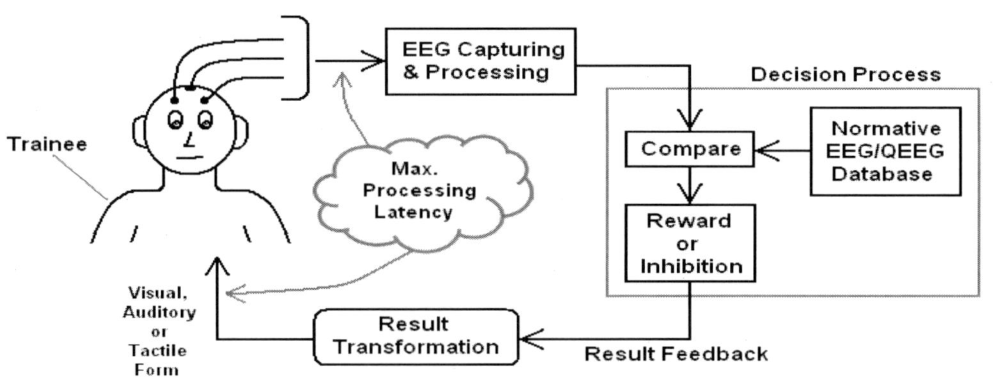

Figure 3. Maximum processing latency of neurofeedback.

STATISTIC INFORMATION

Inclusion/Exclusion Criteria

The objective of inclusion/exclusion criteria is to carefully screen the subjects that comprise a normative database so that representative samples of healthy and otherwise normally functioning individuals are selected, and individuals with a history of neurological problems, psychiatric problems, school failure and other deviant behaviors are excluded [20]. Representative sampling means a demographically balanced sample of different genders, different ethnic backgrounds, different socio-economic status, and different ages that are used to minimize sampling bias [20]. However, as stated before, the inclusion of some active tasks that are cultural specific, e.g. auditory and visual words recognitions, leads to the normative database to become cultural or ethnic specific.

This means that for those active tasks, a separate cognitive activation database is required for one specific language or ethnic group. This may induce a contradiction to the ultimate goal of a normative database stated in [20].

Age Distribution

It is important that a reference normative database contains an adequate sample size per age group, and it spans the age range from birth to adulthood [73]. The adequacy of the sample size will vary depending upon the age under investigation, for example the sample size in young children should be large enough to resolve EEG changes related to their rapid brain development [73]. However, there is no common agreement on how to define the range of ages for each group so that the EEG/QEEG features within a group are similar. Moreover, there is no any statistical information to show how many subjects should be included in a particular age group so that a statistically normal database could be obtained.

Study or Test of Sensitivity, Specificity, Reliability, Gaussian Validation and Clinical Correlation

Currently, the decision of "normal" and "abnormal" is mostly based on "parametric" statistics in which each measurement is independent of all other measures and each one exhibits a histogram shape referred to as a Gaussian or normal distribution [73]. According to [95], for a random variable X which has a normal distribution with mean μ and standard deviation σ, its probability density function (p.d.f.) is defined as:

$$f(x) = \frac{1}{\sqrt{2\pi}\sigma} e^{\left[-\frac{(x-\mu)^2}{2\sigma^2}\right]}$$

This distribution can be transformed to a standard normal distribution that has $\mu = 0$ and $\sigma = 1$ and its p.d.f. becomes:

$$f(z) = \frac{1}{\sqrt{2\pi}} e^{\left(-\frac{z^2}{2}\right)}$$

where $Z = \frac{X-\mu}{\sigma}$. Here the value of Z is called Z-score which is a deviation in number of standard deviation units measured along the base line of the curve from a mean of zero [73]. This Z-score is used to determine whether the measured EEG feature is "normal" (i.e. within a defined range of Z-score) or "abnormal" (i.e. outside the defined range). However, it should be noted that the normal distribution is a hypothetical distribution that is never actually achieved in real-life sampling procedures [73]. The ideal of the normal distribution equation can only be obtained by using some mathematical transformations but, in real life, the ideal normal distribution can never be exactly reproduced [73]. Any deviation from the ideal normal distribution shape will result in different sensitivity, specificity and reliability when using the Z-score method to classify "normal" and "abnormal" brain waves and other QEEG features [73,96]. Therefore, the database provider must minimize the departures from a normal distribution for those data which are stored in a normative database [73,97]. And the final result should be cross-validated to ensure that the deviation from an ideal normal distribution is not critical [97]. More details of these statistic requirements of a normative database can be found in [73], [96] and [97].

Time-of-Day Factors Control

This control refers to the technique used by a normative database provider to handle the time-of-day factors that influence the frequency spectrum of the EEG [97,98]. Two approaches are used. One is statistical randomization. As stated in [97], statistical randomization involves randomly varying time of day of EEG acquisition, time between food intake and EEG acquisition, food content, etc., across ages, sex and demographics. The other method is to obtain sufficient data to allow the analysis of time-of-day effects [98]. For this method, each subject provides two to four replications of each recording condition across several time periods and then these data generate a combination of cross-sectional and longitudinal outcomes over varying time categories to evaluate diurnal influences on EEG characteristics.

Dr. Beverley Steffert (Birkbeck College, London University, UK) had the following comments during a personal communication:

> "There are a number of factors that will affect the EEG spectral components, such as time of day, last food intake, time since last caffeine intake and medications use. Strong emotions like anxiety and anger can also affect EEG. It is general recommended for an EEG to be recorded in the morning before 12:00noon. However, the practical issues of controlling for all of these factors can become impractical. Although most EEG recording software will record the date and time of the session, none of the databases use time-of-day correction in calculating statistical deviation." (Steffert, B., personal communication)

In conclusion, there is no common agreement or guidelines to deal with time-of-date factors while generating a normative EEG/QEEG database, in order to minimize (not eliminate) the influences of those factors.

COMPARISONS OF NORMATIVE EEG/QEEG DATABASES

Four normative EEG/QEEG databases were selected in this chapter: Brain Resource International Database (BRID), Human Brain Institute (HBI) Database, NeuroGuide, and SKIL. The selection criteria were as follows:

1. The database was referenced by some papers or books.
2. The database provider has its official web site to provide up-to-date information about its database.
3. The database provider is still active in development.
4. The database provider has reachable email contact such that a questionnaire can be sent directly to the representative.

The information gathering procedures were:

1. By sending questionnaires to the following organizations:

- Brain Resource Ltd.
  - Database: Brain Resource International Database (BRID)
  - URL: www.brainresource.com and www.brainnet.net
- HBImed AG
  - Database: Human Brain Institute (HBI) Database
  - URL: www.hbimed.com
- Applied Neuroscience, Inc.
  - Database: NeuroGuide
  - URL: www.appliedneuroscience.com
- Sterman-Kaiser Imaging Laboratory
  - Database: SKIL
  - URL: www.skiltopo.com

Questionnaires were sent during the period: 25 Feb 2010 to 3 Mar 2010. By the end of 14 Mar 2010, two responses were received. They were from Sterman-Kaiser Imaging Laboratory and HBImed AG. Their original replies are attached to the Appendix.

2. By gathering related information from their official web sites during the period: 26 Jan 2010 to 14 Mar 2010:
   - BRID: [99,100]
   - NeuroGuide: [101,102,103,104]
3. From their relevant papers:
   - BRID: [105,106,107]
   - NeuroGuide: [97]

The comparisons were categorized into three areas: technical specifications, analysis tools and statistic information. This categorization is similar to the case in considerations of selecting a normative EEG/QEEG database. The results are summarized in the following tables: Table 1 – Technical specifications of different normative databases, Table 2 – Analysis tools available in different normative databases, and Table 3 – Statistic information of different normative databases.

SUMMARY

Technical Specifications

Most database providers used 19 electrodes to record EEG while BRID used 26 electrodes. They used either linked-ears or linked-mastoids montage while capturing EEG and most support re-montage afterward. However, the bandwidth of their resulting EEG is different especially at the setting of higher cut-off frequency. The higher cut-off frequency set in BRID is the highest which is 100 Hz while the lowest one is in NeuroGuide which is 30 Hz only. Another difference is in sampling rate. The sampling rate used in NeuroGuide is the lowest which is 100 Hz while the highest one is 1024 Hz that is used in SKIL. Most database providers used visual inspection to remove artifacts as the basic requirement in artifact removal but automatic mathematic methods were used in HBI, e.g. ICA and slow and fast waves removal by setting thresholds for automatic detection and removal.

Analysis Tools

As stated before, eye-closed and eye-open EEG databases are provided in all databases because these two conditions together with brain waves analysis are the fundamental setup in an EEG analysis and a neurofeedback treatment. The spectral analysis with 1 Hz (single bin) or 0.5 Hz resolution seems to become a common standard and most databases support either resolution. However, as suggested by Dr. David Kaiser (Sterman-Kaiser Imaging Laboratory), shown in Appendix 1, the questionnaire did not ask what type of a window function was applied in spectral analysis. Moreover, the number of elements in a window function (i.e. in one epoch) had not been asked in the questionnaire. These two parameters directly affect the effective frequency resolution and time resolution of the results after calculating FFT. In order to solve this time-frequency resolution tradeoff, wavelet analysis is one of the possible solutions supported by HBI. Wavelet analysis, like FFT, is a mathematical transformation. But wavelet analysis transforms the original signal in terms of "scale" and "translation" of a particular set of signals called "wavelets" [19]. From the results of a wavelet analysis, a set of weights of a particular wavelet that represents a range of frequencies will be shown on a time axis. This means that a continuous change of the amplitude of a particular range of frequencies can be shown in a wavelet analysis. For those reading and learning tasks, BRID and SKIL support English while HBI supports German only. Although BRID does not disclose what analyses are included in the database package, it has the most variety of active tasks or paradigms that were used while generating database data. In contrast, the NeuroGuide is the only one that supports real-time LORETA analysis but it does not contain any active tasks.

Table 1. Technical specifications of different normative databases

	BRID	HBI	NeuroGuide	SKIL
No. of electrodes	26	19	19	19
Montage while capturing	Average of A1 and A2 (mastoid) electrodes sites	Linked ears	Linked ears	Linked ears
Re-montage after capturing	NS	Linked ears, common reference and local reference (Hjorth)	Linked ears, average reference, reference free Current Source Density (CSD), bipolar	Montages that have databases: linked ears, Hjorth Laplacian, Spherical Harmonic Expansion Laplacian and Brodmann area montage Montages that can be displayed without storing as databases: longitudinal bipolar, transverse bipolar, average, average 17 (without frontal poles), vertex reference, symmetry (8 homologue parings), symphony, and a proprietary montage (assume EEG of an entire head or hemisphere is a single signal)
Electrode impedance	<5k Ohm	<5k Ohm	<10k Ohm (most of the impedance were <5k Ohm for all subjects)	<5k Ohm
Amplifier matching	NS	Support	Support	Support
Sampling rate	500 Hz	250 Hz or 500 Hz	100 Hz	1024 Hz; may be down-sampled to 128 Hz
Bit resolution	NS	16 bits	NS	16 bits
Bandwidth	0 – 100 Hz	0.16 – 70 Hz	0.5 – 30 Hz	1 to 38 Hz
SNR of the EEG measuring system	NS	<1.5 μV (noise)	NS	12:1 to 18:1 (21.58 dB to 25.11 dB)
EEG measuring system used	QuickCap, NuAmps and Scan software (from Compumedics NeuroScan)	Mitsar-EEG-201 (from Mitsar, Ltd., Saint Petersburg, Russia)	NS	Lexicor, NeuroNavigator
Artifacts to be removed	Eye movements and other non-EEG like signals	Eye blink artifacts; slow and fast waves	Non-EEG like signals	Muscle (including ocular), equipment artifacts
Artifacts removal techniques	Visually screened by technicians; EOG correction by using a technique based on [108]	Eye blink artifacts – by using ICA; Slow and fast waves – by setting thresholds	Visual inspection	Visual inspection with clipping technique
Signals distortion after artifacts removal	NS	Mild distortion only	NS	Observed

Note: NS – Not Specified.

Table 2. Analysis tools available in different normative databases

	BRID	HBI	NeuroGuide	SKIL
Eye-closed / Eye-open	Both	Both	Both	Both
Reading and learning tasks (Languages included)	Spot the real word, Span of visual memory, Digit span, Verbal interference, Switching of attention, Memory recall and recognition, Word generation, Malingering, Habituation, Auditory oddball, Visual tracking, Go-No go, Visual working memory, Executive maze, Startle, Conscious and subconscious processing of facial emotions (English)	GO/NOGO, emotion face recognition, reading task, auditory, three stimulus odd ball, novelty auditory task (German)	Not support	Reading and mathematics tasks (English)
Time series analysis: waves, ERP, ERD/ERS	Waves and ERP	ERP, ERD/ERS, wavelet	Waves	Waves
Spectral analysis: single bin, brain wave rhythms	NS	Both conventional EEG bands and single bin	Both conventional EEG bands and FFT using 0.5 Hz resolution	Both conventional EEG bands and single bin
Measurements	NS	Voltage amplitude (both in absolute and relative); Spectral power (both in absolute and relative)	Voltage amplitude (in absolute); Spectral power (both in absolute and relative)	Voltage amplitude (both in absolute and relative); Spectral power (both in absolute and relative)
Difference measurement	NS	Support	Support	Support
Ratio measurement	NS	Support	Support	Support
Coherence and phase analysis	NS	Support	Support	Support
Inverse calculation support	NS	LORETA, sLORETA and dipole approximation (off-line calculation)	LORETA (both real-time and off-line calculation)	Pascual-Marqui's SHE (Spherical Harmonic Expansion) and a proprietary inverse solution
Max. processing latency	NS	Almost on-line except those inverse calculations	NS	All analyses are off-line

Note: NS – Not Specified.

Table 3. Statistic information of different normative databases

	BRID	HBI	NeuroGuide	SKIL
Inclusion/Exclusion criteria	Subjects were acquired from over 50 laboratories in the USA, United Kingdom, Holland, South Africa, Israel and Australia. Healthy group excluded a personal history of mental illness, physical brain injury, neurological disorder or other serious medical condition and/or a personal history of sleep disorder, learning disability, drug or alcohol addiction, a personal or family history of ADHD, psychotic or genetic disorder, depressive or anxiety disorder. Another group included subjects with confirmed status as clinical disorder or extreme function: major depressive disorder, ADHD, first episode schizophrenia, post traumatic stress disorder, Alzheimer's disease, traumatic brain injury, sleep apnea, mild cognitive impairment, panic disorder, anorexia nervosa, obesity.	Ethnic composition: Swiss (70%), Russian (15%) and Norwegian (15%) [*1] No history of seizers, no psychiatric or neurological disease as well as no history of such disease in family Average of higher marks in school, successful performance at work	Ethnic composition: whites (71.4%), blacks (24.2%) and oriental (3.2%) 58.9% males and 41.1% females Normalcy for the age range from two months to 18 years was determined by one or more exclusion/inclusion criteria: (1) a neurological history questionnaire given to the child's parents and/or filled out by each subject, (2) psychometric evaluation of IQ and/or school achievement, (3) for children, the teacher and class room performance as determined by school grades and teacher reports and presence of environmental toxins such as lead or cadmium. A neurological questionnaire was obtained from all of the adult subjects more than 18 years of age and those in which information was available about a history of problems as an adult were excluded. The criteria for entry into the normative database for those subjects given IQ tests and school achievement tests were: (1) A Full Scale IQ > 70. (2) WRAT School Achievement Scores > 89 on at least two subtests (i.e. reading, spelling, arithmetic) or demonstrated success in these subjects. (3) A grade point average of 'C' or better in the major academic classes.	Ethnic composition: Caucasians (95%), a few African Americans and a few Asians [*1] High educational levels for adults, 25% female, 75% male, no psychoactive medication, no history of neurological insult or psychiatric diagnosis, and normal sleep the previous day

Table 3. (Continued)

Age distribution	Healthy group: 5000 subjects from the age 6 – 100 Group with disorder: NS	From the age of 7 – 89 Age group 7 – 17: 300 subjects Age group 18 – 60: 500 subjects Age group >60: 200 subjects	From the age of 2 months to 82 years Age group 1 – 13: 470 subjects Age group 14 – 83: 155 subjects (Cf. [103]– Figure 2)	Age groups: 6 – 11 years, 12 – 19 years, 20 – 75 years For future databases, they include 10 – 13 years and 20 – 30 years *(Note: The no. of subjects in each age group has not been specified.)*
Study or test of sensitivity, specificity, reliability, Gaussian validation and clinical correlation	Test of reliability by using "test-retest" method in [106] Cross-cultural assessment in [107].	Not provided	Tested and reported in [103] and [97]	Test-retest capability within the software
Time-of-day factors control	No	No	By randomization	Include time-of-day databases in previous versions of software: 21 one-hour intervals every 30 minutes from 8 am to 5 pm.

Note: NS – Not Specified.

*1 This information was obtained by a follow-up email after receiving the response of the questionnaire.

This means that no database can be used in a real-time analysis of cortical activation while a trainee is performing a cognitive task. For others that support inverse calculations, HBI and SKIL, they can only analyze in off-line mode after the end of EEG recording. Moreover, SKIL can only process and analyze EEG signals in off-line mode. This means that all signal processing and data analyses done by SKIL are executed after recording EEG. For the maximum processing latency, it seems that it has never been measured on a particular machine setup.

Statistic Information

For the inclusion/exclusion criteria, all database providers carefully selected those "normal" or "healthy" subjects before recording normative EEG into their databases. In addition, BRID included another group that was composed of subjects with confirmed status as clinical disorder or extreme function.

For the inclusion of ethnic, only few subjects were Asians and Africans, and all database providers mixed all those subjects into one database. As mentioned before, for those active tasks that are language-related, the database becomes cultural specific and a separate cognitive activation database is required for one specific language or ethnic group. For the age distribution, different databases partitioned the age groups differently. There is still no common agreement on how to partition different age groups such that the EEG/QEEG features within a group are similar. For the sensitivity, specificity, reliability, Gaussian validation and clinical correlation investigation study, BRID and NeuroGuide have done a lot of works while HBI and SKIL have done few only. As stated before, the control of time-of-day factors has been done in two approaches. One is randomization that was used in NeuroGuide and the other is to obtain sufficient data to allow the analysis of time-of-day effects that was used in SKIL. However, at this moment, there are still no studies to show which method is better.

Review of Studies of Neurofeedback for Dyslexia

Up to the end of June 2012, there were only few studies which focused on studying the efficacy of neurofeedback on people with dyslexia. In this section, eight studies of applying neurofeedback to dyslexics are reviewed, in which one of them was a follow-up study that re-examined the learning performance of those subjects in two years after applying neurofeedback treatment to them. Moreover, the strict inclusion of dyslexics only is too narrow. So the studies of those subjects with learning disabilities are also included in this review. This means that this review also includes those studies which examined the efficacy of neurofeedback on people who might also have disorder of mathematics and/or written expression in spite of reading disorder. The design of each experiment, the subject inclusion criteria, the treatment protocols and the results of those eight studies are summarized in Table 4.

For the study of [109], it was a case study that describing the clinical features of children who had their learning disorders successfully treated via EEG neurofeedback training [110].

In the study of [109], 21 males and 3 females, age from 7 years and 4 months to 15 years and 6 months, each with a history of learning disabilities, whose parents chose to have their learning disabilities treated with EEG neurofeedback training were included. Those children were confirmed by the author that all were presenting with learning disabilities. Eleven subjects were primarily diagnosed by their child study team as Neurologically Impaired, eleven as Perceptually Impaired, and two with diagnoses of attention deficit disorder. Their mean WISC-R (Wechsler Intelligence Scale for Children – Revised Edition) IQ scores were as follows: Verbal 98.8 with a standard deviation (SD) of 16.2 and a range of 72 to 128, Performance 98.1 with a SD of 16.9 and a range of 72 to 131, and Full Scale 98.3 with a SD of 16.5 and a range of 70 to 126. Four studies [110,111,112,113] notably showed that the learning disabled had significantly higher amplitudes of brainwave activity in the 7 Hz (theta) band than in the 14 Hz (sensorimotor) band. Therefore, the author chose to apply EEG neurofeedback to increase the amplitude of 14 Hz brainwave rhythm. A single channel Nova Systems "Tansey Ten" Biofeedback EEG was used to assist the subjects in orienting on the 14 Hz (with a 24 dB per octave roll-off filter) neural discharge rhythm over the central Rolandic cortex. In monitoring 14 Hz brainwave production, three Nova Systems saline sensors (with 1k Ohm impedance in saline) were used. The active sensor was placed so that its 6.5 cm × 1.3 cm contact surface laid lengthwise along the midline of the top of the skull (overlaying the cerebral longitudinal fissure), centering about 2.6 cm behind CZ. The reference and ground sensors were randomly placed on opposite ears via comfortable ear clips. The single channel "Tansey Ten" EEG unit provided both amplitude and frequency modulated feedback. The feedback tone was modulated so that the greater the amplitude of the brainwave, the louder the tone. And the repetition rate of the tone (number of beats per second) co-varied with the rate of occurrence of the monitored brainwave energy as it exceeded threshold that was set to 3 μV peak-to-peak. The training was conducted weekly with 40 minutes per session. And the average number of sessions for the study group was 27.9. The results showed that the brainwave signature seemed to normalize and shift about alpha (10 Hz); with those brainwave bands below (5 Hz and 7 Hz) decreasing in monitored energy levels, and those brainwave bands above (12 Hz to 14 Hz) increasing in monitored energy levels. A repeated measures analysis of variance, at each brainwave band, showed that these changes were not due to chance but reflected significant changes in the monitored brainwave signatures. The energy in the lower brainwave bands as monitored (5 Hz and 7 Hz) diminished by 0.088% ($F(1,23)=5.938$; $p<0.05$) and 16.08% ($F(1,23)=11.670$; $p<0.05$) respectively; while the energy in the 10 Hz band did not change significantly ($F(1,23)=0.267$; non-significant) with the 12 Hz and importantly the 14 Hz brainwave bands showed highly significant positive shifts in mean energy by 21.39% ($F(1,23)=18.575$; $p<0.0005$) and 48.95% ($F(1,23)=64.827$; $p<0.0005$) respectively. As observed, the 2.1 times greater amount of energy in the theta band (7 Hz) as compared with the sensorimotor/SMA band (14 Hz) which characterized the pretreatment learning disabled state of affairs was reduced by 64.3% through that EEG neurofeedback regimen. By comparing the WISC-R scores in post- and pre-treatment, 22 of the 24 subjects manifested increases in their Full Scale IQ scores of at least one standard deviation (15 IQ points); with the remaining 2 cases showing an increase of 14 IQ points and 13 IQ points. Both Verbal and Performance IQ scores increased for all subjects. As a group, their post-EEG neurofeedback training scores substantially exceeded their pre-EEG neurofeedback training scores over and above any expected gains due to a retest factor; i.e. more than 3.5 points increase in VIQ, more than 9.5 points in PIQ, and more than 7 points in FSIQ [114]. (Please note that statistical criteria for significance were those provided by [115]).

Another study [116], which did not have a control group and randomization, attempted to replicate the findings of lowered theta/beta ratio resulting from neurofeedback with people with attention deficit disorders and learning disabilities, who also improved on measures of academic achievement. In this study, children and adolescents ages 8 to 18, who were accepted into the study, were referred for academic problems and/or attentional deficits. Initially, forty-three subjects were included in the evaluation and treatment sequence. Each except two was diagnosed with attention deficit disorders by the psychologist author according to DSM-III-R criteria. Two were diagnosed with a developmental visual-motor integration disorder. However, four did not complete the minimum 40 treatment session requirement and the final evaluation and the one who was on Ritalin medication, was changed to other medications and doses during the time of treatment. Then, those five subjects were removed from the study, and finally, 8 females and 30 males participated. This time, the training protocol employed 16-20 Hz (beta) augmentation training with concurrent inhibition of excessive 4-8 Hz (theta) amplitudes because of the findings in the following studies. Wikler, et al. [117] discovered a slow brain wave pattern (theta: 4-8 Hz) dominated in children exhibiting behavior and academic problems. Lubar, et al. [111] assessed a larger number of learning disabled children and a group of normal control children. They found that learning disabled children exhibited slower brain wave patterns than the controls. Mann, et al. [118] showed significantly higher theta and lower beta differences between attention deficit disordered/learning disabled children and normal controls. Other studies, Satterfield and Braley [119] and Linden [120] confirmed the relationship of slow brain wave patterns and attention deficit disorder. During neurofeedback training, a two-channel EEG amplifier from Stolting Autogenics (A620) was used. Electrode placement was bipolar at FZ and PZ and the ground was placed on the left ear. All electrode-skin contact impedance was below 10k Ohms. The training was scheduled at least twice a week but some subjects were trained up to five times per week during summer months. Finally, training was conducted for an average of 46 sessions and median time in treatment was 18 weeks. During the training, the subject was taught to maximize rewards by attending to one of the three displays: a light bulb that increased in size, a fish that followed a course, and a wheel that filled with colors while playing a two octave musical scale. Concurrent with the video display which counted the rewards, an auditory system presented sounds to announce the rewards. A second visible display showed red when the inhibit threshold was exceeded. After completing the treatment, the results showed that all theta/beta ratios and performance measures changed significantly from pre- to post-treatment. The theta/beta ratios under the following conditions declined significantly after treatment: baseline (mean of decrease after treatment: 0.5553; $p=0.001$), eye-closed (mean of decrease after treatment: 0.5403; $p=0.001$), reading (mean of decrease after treatment: 0.4787; $p=0.002$), listening (mean of decrease after treatment: 0.5863; $p=0.000$), drawing (mean of decrease after treatment: 0.6237; $p=0.001$), and concentration (mean of decrease after treatment: 0.4853; $p=0.003$). During the drawing portion, the Beery test of Visual-Motor Integration (BVM) was administered and the results showed that the score increased significantly ($p=0.000$) by 14.974 points in average after treatment. The Jastak Wide Range Reading (WRR), Spelling (WRS), and Arithmetic (WRA) increased significantly by 7.026 ($p=0.000$), 3.947 ($p=0.000$), and 4.184 ($p=0.022$) after treatment respectively.

In the study of [58], it only showed how EEG neurofeedback could help patients who had reading disability and traumatic brain injury by presenting several clinical cases. Moreover, there were not many details of how those subjects were diagnosed with reading disorder and what treatment protocols were used. But that study demonstrated how to use active tasks intervention in a neurofeedback treatment in order to improve reading and memorizing performance. One of the authors in [58] (K. E. Thornton) developed a QEEG activation database with normal child and adult subjects (US Patent No.: 6309361). An activation database is necessary because the eye-closed condition does not provide information on the brain's active functioning [58] Thornton and Carmody [58] suggested the following treatment protocols and intervention methods to improve reading and memorizing performance of patients with reading disorder or traumatic brain injury. The treatment consisted of subjects either listening to audiotapes or reading while the appropriate protocols were being used. The purpose of this approach was to train the brain under the appropriate and relevant task conditions. During treatment, the subject's progress was tested with novel stories that contained approximately 20 to 25 pieces of information. The subject listened to the story at the beginning of the session and recalled the story immediately to the clinician to obtain an immediate memory score. At the end of the session, the subject was asked to recall the story to obtain a delayed memory score. This treatment was applied to subjects and two cases were described in [58]. One case was a boy with significant history of reading problems. The treatment was directed toward the posterior connection problem of Beta 1 (12-32 Hz) and Beta 2 (32-64 Hz) from the occipital positions under reading and auditory memory conditions. His auditory memory functioning increased 589% from baseline by the end of the twenty-fifth session. He also improved on the standardized reading inventory (SRI) from the previous year's testing from a Lexile score of 360 before treatment to 753 after approximately 40 sessions, which was a much larger change than the typical improvement of 75 to 100 Lexiles per year. Moreover, the deviation of Beta 1 and Beta 2 coherence and phase at occipital positions between the subject and the data in Thornton's QEEG activation normative database reduced (i.e. shifted toward normal). Another case was a 17-year-old boy with reading disability. After 20 sessions, he increased his comprehension score (on the Burns Roe Reading Passages) from 45% to 90% (on alternate versions – eighth grade level) and from 20% to 70% (on tenth grade level). His performance on alternate versions of the Cognisys Story Recall Test had increased by 3 standard deviations. On the Wechsler Individual Achievement Test reading comprehension subtest he attained a standard score of 99 for age and grade level.

In the study of [2], twelve clinical cases of using neurofeedback to help dyslexic individuals to learn to read more easily were reviewed. And all subjects were patients in the authors' clinic. In this study, the treatment protocols were different for all cases (see Table 4). First of all, they obtained a QEEG and a reading difference topograph from a patient. Then they trained down any abnormalities that were significantly increased and trained up any abnormalities that were significantly decreased. Those treatments were done either in eye-open condition or under reading condition. This meant that the authors of this study also used active tasks intervention to train their patients like the cases in [58], but the reading tasks used in those two studies were different. After neurofeedback training, each of the twelve individuals improved by at least two grade levels after 30 to 35 sessions.

In the study of [79], eight boys and two girls were included to study the association between the behavioral improvement and EEG modification in those learning disabled children after neurofeedback treatment. Those subjects were selected according to the inclusion criteria listed in Table 4. Some of the inclusion criteria should be emphasized here. All subjects did not have other neurological or psychiatric disorders except for the presence of "Learning disorder not otherwise specified" according to DSM-IV criteria. This meant that those subjects not only had "reading disorder" (dyslexia), but also might have "mathematics disorder" and/or "disorder of written expression" that were defined in DSM-IV. Moreover, they also excluded those subjects who were diagnosed with ADHD (Attention-Deficit/Hyperactivity Disorder). Finally, they only included those learning disabled children with abnormally high value of theta/alpha ratio in eye-closed condition when compared to a normative database. Those subjects were divided into the experimental group and the control group. The experimental group received "real" neurofeedback treatment while the control group received "placebo", in which all conditions were exactly the same as in the experimental group except that the reward and its duration were random and non-contingent with EEG activity. The ethical use of placebo controls in clinical research has been the subject of debate [79]. But Fernández, et al. [79] stated that, according to the Helsinki Declaration, placebo-controlled studies are ethically acceptable for those disorders for which no effective treatment is available [121]. For the treatment protocol, neurofeedback was conducted using a neurofeedback program adapted to the MEDICID IV recording system and software. EEG recordings were obtained from a monopolar lead, situated at the place with the most abnormal theta/alpha ratio with linked earlobes as the reference. The training targeted to decrease the theta/alpha ratio, (theta: 3.6-7.5 Hz; alpha: 7.6-12.5 Hz) at that selected place. This was based on that several studies reported higher amounts of delta, and principally theta, and lower amounts of alpha in learning disabled children compared to normal children of the same age [122, 123, 124] and a minimum quantity of alpha absolute power at rest was necessary for the correct performance of mental tasks [125,126]. Each training session was 30 minutes long and totally 20 sessions were conducted over a period of 10-12 weeks. The results showed that the experimental group had significant increase in ADHD score from TOVA (Test of Variables of Attention) and in the following WISC-R scores: Total IQ ($p=0.02$), Verbal ($p=0.03$), and Performance ($p=0.04$). But the control group did not show any significant improvement in those tests. For the EEG results, no significant difference was found between the EEG recorded before treatment and the EEG recorded immediately after treatment finished. However, EEG changes were observed in 2 months after neurofeedback treatment. In two months after completing the treatment, in the test of the global hypothesis, significant absolute power reduction in all frequency bands was found in the experimental group ($p<0.05$), whereas in the control group, only the reduction in relative power of delta band (0.5-3.5 Hz) was significant ($p<0.05$). When marginal hypotheses were tested, significant changes ($p<0.05$) in many leads were observed in the experimental group.

The study of [127] was a follow-up study that was carried out two years after the end of the neurofeedback or placebo treatment done in [79]. Its aim was to analyze the EEG and behavioral changes that had occurred in two years after the end of treatment. This study included the same subjects who participated in [79], except that one child in the control group left the school one year prior to the follow-up study and declined to participate. The results showed that, in the experimental group, WISC-R scores in the global IQ ($p=0.04$) and the Performance ($p<0.05$) increased significantly two years after neurofeedback treatment, but the Verbal scores decreased in the last two years with respect to both before and immediately after neurofeedback treatment. In the experimental group, the ADHD score from TOVA also increased significantly two years after as compared to the scores both before and immediately after neurofeedback treatment. In contrast, the control group did not show significant difference between comparisons any time in WISC-R or in ADHD scores from TOVA. The most important result obtained in this study was that two years after treatment, the children in the control group continued presenting learning disorder, but four out of five children in the experimental group had a normal diagnosis. This meant that four out of five children had overcome their learning disorder after the neurofeedback treatment.

Another study [80] was directed by the same organization that directed the study of [79]. The study of [80]. was also a placebo-controlled study of neurofeedback for learning disabled children. This time, the experimental protocol was approved by The Ethics Committee of Instituto de Neurobiología, Universidad Nacional Autónoma de México. In this study, nine boys and seven girls were referred by a social worker from several elementary schools in Querétaro, Mexico. The authors used the same inclusion criteria and neurofeedback protocols as that in [79]. The results showed that, in the control group, no significant difference was observed in WISC-R scores before and after the treatment. In the experimental group, WISC-R scores in the IQ ($p=0.02$) and Verbal ($p=0.001$) increased significantly, but the Performance did not. RM-ANOVA analyses revealed no significant difference between groups for IQ, Verbal, or Performance score. The ADHD score from TOVA was invalidated for one child in the experimental group. In the experimental group, significant increases ($p=0.01$) in the ADHD score from TOVA were observed. When reaction time, false alarm percentage, and omission percentage were compared before and after neurofeedback treatment, a highly significant ($p=0.001$) decrease in omission percentage was observed in the experimental group. The RM-ANOVA results were highly significant only for the interaction group across condition ($F(1,13)=10.937$; $p=0.006$). All children of the experimental group learned to decrease the theta/alpha (theta: 3.6-7.5 Hz; alpha: 7.6-12.5 Hz) ratio during the course of the neurofeedback sessions. This learning was not so evident when the comparison between EEG at rest recorded before and immediately after treatment was made. But, two months later, more children showed a decrement of this ratio. For the EEG current sources that were calculated by using FD-VARETA (Frequency Domain Variable Resolution Electromagnetic Tomography), in the control group, no significant difference was observed. In the experimental group, all current decreases occurred in the delta (1.5-3.5 Hz) and theta (3.6-7.5 Hz) bands, and all current increases occurred in the alpha (7.6-12.5 Hz) and beta (12.6-19 Hz) bands.

Two months later, many changes were found. A current reduction was found mainly in the frequencies within the theta band, in the cingulate region, temporal lobe, fronto-orbital gyrus of the left hemisphere, and in the right parahippocampal area. There were current increases in frequencies within the alpha band in temporal and frontal regions of the right hemisphere, and in frequencies within the beta band in frontal, temporal, and cingulate areas. With the objective of formalizing the comparison between groups across time, an RM-ANOVA test was performed, considering condition as the within factor, and group as the between factor. No significant difference was observed in any factor or interaction. These results should be taken with caution, since the assumption about the normal distribution of the data was not fulfilled.

The most recent study [128] was done in Netherlands which attempted to reduce reading and spelling deficits in children who were dyslexic by means of neurofeedback training based on neurophysiological differences between the participants, and gender and age matched controls. This study included eleven boys and eight girls who were diagnosed with dyslexia by their remedial teachers based on a structured protocol assessing the reading and spelling development of the children [129] and all of them were attending regular schools. They were also screened such that they did not have any properties that were listed in the exclusion criteria (see Table 4).

This study used a randomized controlled pretest-posttest design. This time, the experimental group received neurofeedback treatment while the control group did not. But both groups received additional counseling regarding reading and spelling, apart from the regular language curriculum. The neurofeedback protocols used in this study were based on the outcome of the QEEG assessment of each subject. In addition, coherences were checked and compared to the outcome of the study in [43]. The details of treatment protocols for each subject are listed in Table 4. All participants were given 20 sessions of neurofeedback training during 10 weeks. The children were administered a range of tests to further investigate correlations between EEG and neuropsychological findings to sub-tests of dyslexia. The included tests were measures of tasks related to reading-rapid naming of letters, articulation, phoneme deletion [130] and spelling [131]

Table 4. Summary of studies of neurofeedback for dyslexia

Tansey [109]	
Design	No control; no randomization; case study
Subjects	• 21 males and 3 females • Age: 7 years 4 months to 15 years 6 months • Each with a history of learning disabilities (the learning disabilities were confirmed by the author) whose parents chose to have their learning disabilities treated with EEG neurofeedback training • 11 subjects were diagnosed as Neurological Impaired, 11 subjects were diagnosed as Perceptually Impaired and 2 subjects were diagnosed with attention deficit disorder by their child study team. • Their mean WISC-R (Wechsler Intelligence Scale for Children – Revised Edition) IQ scores were as follows: Verbal 98.8 with a standard deviation (SD) of 16.2 and a range of 72 to 128, Performance 98.1 with a SD of 16.9 and a range of 72 to 131, and Full Scale 98.3 with a SD of 16.5 and a range of 70 to 126
Protocols	• Electrode placement: 3 saline sensors were placed so that 6.5 cm × 1.3 cm contact surface laid lengthwise along the midline of the top of the skull (overlaying the cerebral longitudinal fissure), centering about 2.6 cm behind CZ (International 10-20 System). The reference and ground nodes were randomly placed on opposite ears. • Target frequencies: Increase 14 Hz (i.e. Increase the amplitude of frequency band around 14 Hz) brainwave • Training schedule: 40 minutes per session; scheduled once a week; average no. of sessions: 27.9
Results	• Brainwave: The brainwave bands at 5 Hz and 7 Hz decreased and those brainwave bands at 12 Hz and 14 Hz increased in monitored energy levels. • WISC-R: Full Scale, Verbal and Performance IQ scores were assessed. 22 of the 24 subjects manifested increases in their Full Scale IQ scores of at least one standard deviation (15 IQ points), with the remaining 2 cases showing an increase of 14 IQ points and 13 IQ points. Both Verbal and Performance IQ scores increased for all subjects.
Fenger [116]	
Design	No control; no randomization
Subjects	• 30 males and 8 females • Age: 8 to 18

	• They were referred for academic problems and/or attention deficits.
Protocols	• Electrode placement: Bipolar measurement at FZ and PZ (International 10-20 System); Ground on left ear • Target frequencies: Increase 16-20 Hz with concurrent decrease of 4-8 Hz (i.e. Increase the amplitude of frequencies within 16-20 Hz and decrease the amplitude of frequencies within 4-8 Hz concurrently.) • Training schedule: 45 minutes per session; scheduled at least twice a week but some were trained up to five times per week during summer months; average no. of sessions: 46
Results	• Brainwave: Theta/Beta ratio (TBR) decreased by 15% in average • Beery test of Visual-Motor Integration (BVM): increased by 14.974 points in average • Jastak Wide Range Reading (WRR): increased by 7.026 points in average • Jastak Wide Range Spelling (WRS): increased by 3.947 points in average • Jastak Wide Range Arithmetic (WRA): increased by 4.184 points in average
Thornton and Carmody [58]	
Design	Case study; neurofeedback with active tasks intervention
Subjects	• Subject 1: (It was called Case 3 in [58].) A boy (age was not disclosed) with significant history of reading problems • Subject 2: (It was called Case 4 in [58].) A 17-year-old boy with reading disability
Protocols	• Active Tasks Intervention: The treatment consisted of subjects either listening to audiotapes or reading while the appropriate protocols were being used. During treatment, the subject's progress was tested with novel stories that contained approximately 20 to 25 pieces of information. The subject listened to the story at the beginning of the session and recalled the story immediately to the clinician to obtain an immediate memory score. At the end of the session, the subject was asked to recall the story to obtain a delayed memory score. <u>Subject 1</u> • Electrode placement: Not disclosed • Target frequencies: Directed toward Beta 1 (12-32 Hz) and Beta 2 (32-64 Hz) coherence and phase problems at occipital positions • Training schedule: 40 sessions; (The duration and how often of training sessions – Not disclosed); under reading and auditory memory conditions (i.e. trained with active tasks) <u>Subject 2</u> • Electrode placement: Not disclosed • Target frequencies: Not disclosed • Training schedule: 20 sessions; (The duration and frequency of training sessions – Not disclosed); using activation approach (i.e. trained with active tasks)
Results	<u>Subject 1</u> • Brainwave: The deviation of Beta 1 and Beta 2 coherence and phase at occipital positions between the subject and the data in Thornton's QEEG activation normative database reduced. (i.e. Those brainwaves shifted towards normal.) • Auditory memory functioning increased 589% from the baseline by the end of the 25-th training session. • Improved on the standardized reading inventory (SRI) from a Lexile score of 360 before treatment to 753 after approximately 40 training sessions. <u>Subject 2</u> • Increased his comprehension score (on the Burns Roe Reading Passages) from 45% to 90% (on alternative version – eighth grade level) and from 20% to 70% (on tenth grade level). • His performance on alternative versions of the Cognisys Story Recall Test had increased by 3 standard deviations. • On the Wechsler Individual Achievement Test reading comprehension subtest, he attained a standard score of 99 for age and grade level.

Walker and Norman [2]	
Design	Case study
Subjects	*(Please note that all subjects were patients in authors' clinic.)* • Case 1: A 15-year-old boy presented with a chief complaint of dyslexia; complained of short-term memory difficulty and intermittent depression; studied in the ninth grade, but reading at a fifth grade level • Case 2: A 9-year-old girl had the chief complaints of dyslexia, spelling difficulty and bad handwriting; studied in the fourth grade, but reading at first grade level • Case 3: age: 16; studied in the tenth grade, but reading at ninth grade level • Case 4: age 10; studied in the fourth grade, but reading at first grade level • Case 5: age 11; studied in the fifth grade, but reading at second grade level • Case 6: age 7; studied in the first grade, but reading at pre-K grade level • Case 7: age 11; studied in the fifth grade, but reading at fourth grade level • Case 8: age 12; studied in the sixth grade, but reading at fifth grade level • Case 9: age 9; studied in the third grade, but reading at second grade level • Case 10: age 10; studied in the fifth grade, but reading at fourth grade level • Case 11: age 10; studied in the third grade, but reading at second grade level • Case 12: age 7; studied in the second grade, but reading at first grade level
Protocols	*(Please note that how often the training sessions held was not disclosed.)* <u>Case 1</u> 10 minutes per session; 30 sessions with 5 sessions of each of the following protocols: • Decrease 1-7 Hz at CZ in eyes open condition • Decrease 2-7 Hz and increase 15-18 Hz at CZ in eyes open condition • Decrease 1-10 Hz at O1 while reading • Increase 11-12 Hz at O1 in eyes open condition • Decrease coherence of delta at F8/T6 in eyes open condition • Decrease coherence of delta at T4/O2 in eyes open condition <u>Case 2</u> 10 minutes per session; 53 sessions as follows: • (10 sessions) Decrease 2-8 Hz and increase 15-18 Hz at CZ in eyes open condition • (10 sessions) Decrease 2-8 Hz and increase 15-18 Hz at P3 in eyes open condition • (10 sessions) Decrease 1-10 Hz at O1 while reading • (5 sessions) Decrease coherence of alpha at T4/T6 in eyes open condition • (5 sessions) Decrease coherence of beta at CZ/PZ in eyes open condition • (5 sessions) Decrease coherence of delta at FP1/F7 in eyes open condition • (8 sessions) Increase 11-12 Hz at O2 in eyes open condition <u>Case 3</u> 10 minutes per session; 35 sessions with 5 sessions of each of the following protocols: • Decrease 2-7 Hz and increase 12-15 Hz at FP2 in eyes open condition • Decrease 1-8 Hz and decrease 18-30 Hz at OZ in eyes open condition • Decrease coherence of beta at P3/O1 in eyes open condition • Decrease coherence of beta at FP2/O2 in eyes open condition • Increase coherence of delta at F3/O1 in eyes open condition • Increase coherence of theta at C4/P4 in eyes open condition • Increase coherence of delta at F4/O2 in eyes open condition <u>Case 4</u> 10 minutes per session; 25 sessions with 5 sessions of each of the following protocols: • Decrease 2-7 Hz and increase 15-18 Hz at FP1 in eyes open condition • Decrease 1-5 Hz at O2 in eyes open condition • Decrease 1-8 Hz at PZ while reading • Increase coherence of beta at F4/O2 in eyes open condition • Decrease coherence of theta at F3/P3 in eyes open condition

	Case 5 10 minutes per session; 20 sessions with 5 sessions of each of the following protocols: • Decrease 1-4 Hz at O2 in eyes open condition • Decrease 1-5 Hz at F8 in eyes open condition • Increase 15-18 Hz at T3 while reading • Decrease 10-12 Hz at O1 in eyes open condition Case 6 10 minutes per session; 30 sessions with 5 sessions of each of the following protocols: • Decrease 4 Hz at F7 in eyes open condition • Decrease 5-6 Hz at F8 in eyes open condition • Decrease 29-30 Hz at FZ in eyes open condition • Decrease coherence of beta at O1/T3 in eyes open condition • Decrease 1-10 Hz at F3 while reading • Decrease 2-7 Hz and increase 15-18 Hz at C3 in eyes open condition Case 7 10 minutes per session; 40 sessions with 5 sessions of each of the following protocols: • Decrease 2-7 Hz at F3 and F4 in eyes open condition • Decrease 2-7 Hz at O1 and O2 in eyes open condition • Increase coherence of beta at F7/F4 in eyes open condition • Increase coherence of theta at C4/T4 in eyes open condition • Decrease coherence of delta at O1/F3 in eyes open condition • Decrease 2-7 Hz and increase 15-18 Hz at PZ in eyes open condition • Increase coherence of theta at PZ/O2 in eyes open condition • Decrease coherence of theta at T3/C3 in eyes open condition Case 8 10 minutes per session; 15 sessions with 5 sessions of each of the following protocols: • Decrease 2-7 Hz and increase 15-18 Hz at T5 in eyes open condition • Decrease 2-7 Hz and increase 15-18 Hz at O2 in eyes open condition • Decrease 2-7 Hz and increase 15-18 Hz at O1 in eyes open condition Case 9 10 minutes per session; 20 sessions with 5 sessions of each of the following protocols: • Decrease 2-7 Hz and increase 12-15 Hz at C4 in eyes open condition • Decrease 2-7 Hz and increase 15-18 Hz at O2 in eyes open condition • Decrease 2-7 Hz and increase 15-18 Hz at O1 in eyes open condition • Decrease 2-7 Hz and increase 15-18 Hz at P4 in eyes open condition Case 10 10 minutes per session; 35 sessions with 5 sessions of each of the following protocols: • Decrease 1-4 Hz at CZ in eyes open condition • Decrease 1-2 Hz and increase 9-11 Hz at O1 in eyes open condition • Decrease 1-11 Hz at FP1 while reading • Increase coherence of beta at T3/T4 in eyes open condition • Decrease coherence of alpha at F1/T3 in eyes open condition • Decrease coherence of delta at T5/P3 in eyes open condition • Increase coherence of theta at C3/T3 in eyes open condition Case 11 10 minutes per session; 35 sessions with 5 sessions of each of the following protocols: • Decrease 2-7 Hz and increase 15-18 Hz at C3 in eyes open condition • Decrease 1 Hz and increase 15-18 Hz at T4 in eyes open condition • Decrease 1-10 Hz at F3 while reading • Decrease 30 Hz at F8 in eyes open condition

	• Decrease coherence of delta at F7/T5 in eyes open condition • Decrease coherence of alpha at O1/T3 in eyes open condition • Decrease coherence of theta at F3/T5 in eyes open condition <u>Case 12</u> 10 minutes per session; 25 sessions with 5 sessions of each of the following protocols: • Decrease 1-8 Hz and increase 15-18 Hz at CZ in eyes open condition • Decrease coherence of alpha at F7/F3 in eyes open condition • Increase coherence of delta at CZ/PZ in eyes open condition • Increase coherence of alpha at CZ/C4 in eyes open condition • Increase 11-12 Hz at O2 in eyes open condition
Results	<u>Case 1</u> • Brainwave: QEEG showed normalization of the reading difference topography with a decrease in frequency range of 1 to 10 Hz in the reading condition. The coherence abnormalities were normalized. • Reading grade level advanced to Grade 10; no longer having memory problems and depressed by self-report <u>Case 2</u> • Brainwave: Reading difference topographies revealed the normal decrease in 9 to 10 Hz activity in the reading condition, though the occipital 4 to 7 Hz frequencies and 1 to 2 Hz frequencies at F7 and F8 were still increased during reading (better, but not normal). • Reading level advanced to Grade 6, as judged by her reading teacher. She was reading for pleasure for the first time. Her handwriting and spelling were much improved. <u>Case 3</u> • Reading grade level advanced from Grade 9 to 12 <u>Case 4</u> • Reading grade level advanced from Grade 1 to 4 <u>Case 5</u> • Reading grade level advanced from Grade 2 to 5 <u>Case 6</u> • Reading grade level advanced from Grade Pre-K to 2 <u>Case 7</u> • Reading grade level advanced from Grade 4 to 10 <u>Case 8</u> • Reading grade level advanced from Grade 5 to 7 <u>Case 9</u> • Reading grade level advanced from Grade 2 to 4 <u>Case 10</u> • Reading grade level advanced from Grade 4 to 7 <u>Case 11</u> • Reading grade level advanced from Grade 2 to 4 <u>Case 12</u> • Reading grade level advanced from Grade 1 to 3
Fernández, et al. [79] and Becerra, et al. [127]	

Design	The study of [79] was a controlled study (by applying placebo to the control group) but without randomization. Two years after completing the study of [79], another study [127] was done to re-examine the learning performance of all subjects that were studied in [79].

Subjects	Experiment done in [79]
	• 8 males and 2 females from Mexico
	• Age: 7 to 11
	• Inclusion Criteria:
	  • All were diagnosed with "Learning disorder not otherwise specified" according to DSM-IV criteria.
	  • mother with at least a third-grade elementary school education
	  • per capita income greater than 50% of minimum wage
	  • no neurological or psychiatric disorders except for the presence of "Learning disorder not otherwise specified" according to DSM-IV criteria
	  • no important alterations in their brain Computed Tomography
	  • their IQ scores were at least 70
	  • they did not have severe sociocultural disadvantages
	  • all had an abnormally high EEG theta/alpha ratio at least in one electrode when their ratio at rest with eye closed was compared with a normative database for their age
	  • no paroxysmal activity in the alpha frequency range
	  • not satisfy the criteria to be classified as ADHD according to DSM-IV
	  • All were volunteers.
	  • Experimental Group: 4 males and 1 female receiving neurofeedback treatment
	  • Control Group: 4 males and 1 female receiving placebo treatment
	
	Experiment done in [127]
	• Experimental Group: The same Experimental Group in [79]; age: 9.08 to 12.58
	• Control Group: 4 of the 5 subjects in the Control Group in [79]; age: 10.33 to 14.33; (The other child, who had received placebo treatment, left the school one year prior to the follow-up study and declined to participate in the tests.)
	• No significant difference between groups existed in age, IQ, and ADHD score from Test of Variables of Attention (TOVA) in any of the three time periods: before, after and 2 years after neurofeedback treatment.

Protocols	Experiment done in [79] Experimental Group: • Electrode placement: An electrode was placed at the place (according to the International 10-20 System) with the most abnormal theta/alpha ratio (power ratio); used linked earlobes as a reference (Theta: 3.6-7.5 Hz; Alpha: 7.6-12.5 Hz) • Target frequencies: Decrease the theta/alpha ratio (power ratio) at the selected electrode • Training schedule: 30 minutes per session; total 20 sessions over a period of 10-12 weeks Control Group: • The Control Group received the placebo treatment, in which all conditions were exactly the same as in Experimental Group, except that the reward and its duration were random, non-contingent with EEG activity.
Results	Experiment done in [79] Experimental Group: • EEG absolute power decreased in delta (0.5-3.5 Hz), theta (3.6-7.5 Hz), alpha (7.6-12.5 Hz) and beta (12.6-19 Hz) bands • Significant improvement in WISC-R scores in Total IQ, Verbal and Performance • Significant improvement in ADHD score from TOVA Control Group: • EEG relative power decreased in delta band only • No significant improvement in WISC-R scores and ADHD score from TOVA Experiment done in [127] Experimental Group:

	• The global IQ and the Performance scores from WISC-R increased significantly, but the Verbal score in WISC-R decreased • The ADHD score from TOVA increased significantly (i.e. improved) • 3 out of 5 children improved their self-esteem; 4 improved their attention and behavior; all children improved their memory, school attitude and scholastic achievement • 4 out of 5 children had overcome learning disabilities Control Group: • No significant difference between comparisons any time in WISC-R or in ADHD scores from TOVA • Only 3 out of 4 children who wanted to participate in the study improved their attention, self-esteem, behavior, school attitude and scholastic achievement and only 2 improved their memory in the opinion of their parents; one child did not show any change • All continued presenting learning disabilities
Fernández, et al. [80]	
Design	Controlled study (by applying placebo to the control group); no randomization
Subjects	• 9 males and 7 females from Mexico • Age: 7 to 11 • They were referred by a social worker from several elementary schools in Querétaro • Inclusion Criteria: • All were diagnosed with "Learning disorder not otherwise specified" according to DSM-IV criteria by using the tests that were the arithmetic subscale of the WISC-R, and a writing-reading test standardized by grade for Mexican children [134] • presented learning disabilities with normal neurological exam • WISC-R IQ score greater than 70 • mother with at least a third-grade elementary school education • per capita income greater than 50% of the minimum wage • an EEG recorded at rest with eyes closed in which at least one lead showed an abnormally high value of the theta/alpha ratio compared to a normative database • without paroxysmal activity in the alpha frequency range • not diagnosed with ADHD or other psychiatric alteration according to DSM-IV criteria • without major brain abnormalities found in cranial Computed Tomography • All were volunteers. • Experimental Group: 5 males and 6 females with age from 7.16 to 10.68 receiving neurofeedback treatment • Control Group: 4 males and 1 female with age from 7.83 to 11.36 receiving placebo treatment
Protocols	Experimental Group: • Electrode placement: An electrode was placed at the place (according to the International 10-20 System) with the most abnormal theta/alpha ratio (power ratio); used linked earlobes as a reference (Theta: 3.6-7.5 Hz; Alpha: 7.6-12.5 Hz) • Target frequencies: Decrease the theta/alpha ratio (power ratio) at the selected electrode in eye-closed condition • Training schedule: 30 minutes per session; total 20 sessions over a period of 10-12 weeks Control Group: • The Control Group received the placebo treatment, in which all conditions were exactly the same as in Experimental Group, except that the tone onset (reward) and its duration were random, non-contingent with EEG activity.
Results	Experimental Group: • Brainwave: All children learned to decrease the theta/alpha ratio. Current sources that were calculated by Frequency Domain Variable Resolution Electromagnetic Tomography (FD-VARETA), showed decreases in the delta (1.5-3.5 Hz) and theta (3.6-7.5 Hz) bands, and all current increases occurred in the alpha (7.6-12.5 Hz) and beta (12.6-19 Hz) bands. Two months later, a current reduction was found mainly in the frequencies within the theta band, in the cingulate region, temporal lobe, fronto-

	orbital gyrus of the left hemisphere, and in the right parahippocampal area. There were current increases in frequencies within the alpha band in temporal and frontal regions of the right hemisphere, and in frequencies within the beta band in frontal, temporal, and cingulate areas. • The IQ and Verbal scores from WISC-R increased significantly but the Performance score did not. • The ADHD score from TOVA increased (i.e. improved) significantly except one child was invalidated. • All parents of children reported behavioral improvement in memory, attention, and attitude toward school. All the parents also reported highly significant positive changes in their children's socialization, self-confidence, and independence behaviors, with one exception. Control Group: • Brainwave: Current sources that were calculated by FD-VARETA did not show any significant difference. • No significant difference was observed in WISC-R IQ, Verbal, and Performance scores. • The ADHD score from TOVA of four out of five children showed deterioration by more than one point (i.e. worse than before treatment). The remaining child improved by less than one point. • According to the parents, two children presented no improvement. The parents of the remaining reported behavioral improvements in memory, attention, and attitude toward school. Only one girl showed an improvement in academic performance.
Breteler, et al. [128]	
Design	Randomized controlled pretest-posttest design

Subjects	• 11 males and 8 females from Netherlands
	• Age: 8.0 to 15.98 (average: 10.33)
	• Inclusion Criteria:
	  • They were diagnosed with dyslexia by their remedial teachers based on a structured protocol assessing the reading and spelling development of the children from grade 1 [129].
	  • The remedial teachers selected those children, who all were attending regular schools.
	  • Exclusion Criteria:
	  • a personal or family history of mental illness
	  • brain injury
	  • neurological disorder
	  • serious medical condition
	  • drug/alcohol addiction
	  • a family history of genetic disorder
	  • All were volunteers.
	  • Experimental Group: 10 subjects; received neurofeedback treatment
	  • Control Group: 9 subjects; no neurofeedback treatment

Protocols	*Please note the following:* • Frequency bands definitions: delta (1.5-3.5 Hz), theta (4-7.5 Hz), alpha (8-13 Hz), alpha1 (8-11 Hz), alpha2 (11-13 Hz), SMR (12-15 Hz), beta (14.5-30 Hz), beta1 (14.5-20 Hz), beta2 (20-25 Hz) and beta3 (25-30 Hz) • "Decrease delta" means to decrease the average power of delta frequency band • "Increase alpha" means to increase the average power of alpha frequency band • Only the Experimental Group received neurofeedback treatment • Both the Experimental and the Control Groups received additional counseling regarding reading and spelling, apart from the regular language curriculum Experimental Group: • Electrode placement: Electrodes were placed according to the outcome of the QEEG assessment of each subject (details listed below); placed according to the International 10-20 System; linked mastoids as the reference with ground at AFZ

	• Target frequencies: Different subjects had different target frequencies (details listed below) • Training schedule: 20 sessions trained during 10 weeks <u>Case 1</u> • Decrease coherence of delta at T3/T4 in eyes open condition <u>Case 2</u> • Decrease 2-5 Hz and increase 15-20 Hz at T6 in eyes closed condition • Decrease coherence of beta at F7/C3 in eyes closed condition <u>Case 3</u> • Decrease coherence of delta at T3/T4 in eyes open condition • Decrease coherence of beta at F7/C3 in eyes open condition <u>Case 4</u> • Decrease 2-12 Hz at T6 in eyes open condition • Decrease coherence of alpha at F7/FC3 in eyes open condition <u>Case 5</u> • Decrease 2-8 Hz at T4 in eyes open condition • Decrease coherence of delta at T3/T4 in eyes open condition • Decrease coherence of alpha at F7/FC3 in eyes open condition <u>Case 6</u> • Decrease coherence of delta at T3/T4 in eyes open condition • Decrease coherence of alpha at F7/C3 in eyes open condition <u>Case 7</u> • Decrease 2-5 Hz and increase beta at T6 in eyes open condition • Decrease coherence of delta at T3/T4 in eyes open condition • Decrease coherence of beta at F7/C3 in eyes open condition <u>Case 8</u> • Decrease 18-20 Hz and decrease 5-8 Hz at FZ in eyes closed condition • Increase 12-15 Hz at C3 in eyes open condition <u>Case 9</u> • Decrease coherence of beta at F7/C3 in eyes open condition <u>Case 10</u> • Decrease 2-4 Hz at F3 • Decrease coherence of delta at T3/T4 in eyes open condition • Decrease coherence of alpha at F7/FC3 in eyes open condition
Results	• Brainwave: There was no change in the delta band at any location. At CZ and P3, theta increased among all participants. Alpha1 was increased at CP4 and at O1 among all participants. Alpha1 was intended to go down at locations F3 and T6. Beta1 decreased at OZ among all subjects. In the coherence of alpha band, a consistent change was found at CP3/F3 in all participants. Other differences in coherence occurred at P3/F3, O1/F3 and C4/T4, showed insignificant correlations. A multitude of coherences went down in the beta range, only P4/T4 showed a consistent decrease among all subjects. • Language tests which included measures of tasks related to reading-rapid naming of letters, articulation, phoneme deletion [130] and spelling [131] were assessed. • The main result was a small (corrected in [132]) and clinically relevant improvement in spelling, whereas no improvement in reading abilities was found.

The results from reading and spelling tests showed that the analysis of variance indicated main effects of the factor "time" for the variables CVC-words ($F(1,17)=11.1$; $p=0.004$), CC-words ($F(1,17)=8.5$; $p=0.01$), double syllables ($F(1,17)=21.4$; $p=0.000$), triple syllables ($F(1,17)=23.7$; $p=0.000$), and CC-pseudo words ($F(1,17)=9.7$; $p=0.006$). No interaction effect was found, suggesting comparable changes in both the Experimental and control groups. The spelling test showed a main effect of time ($F(1,17)=10.2$; $p=0.001$). Additionally, a significant interaction effect was found ($F(1,17)=4.5$; $p=0.045$), suggesting a treatment effect of neurofeedback. The experimental group progressed from m=69.1 (SD=32.0) to m=80.6 (SD=32.2) a 16.6% improvement, whereas the control group progressed from m=66.9 (SD=20.9) to m=70.9 (SD=24.4) a 6% improvement. Cohen's d was 0.26 (this figure was corrected in [132]), suggesting a small and significant improvement in spelling after neurofeedback treatment. But no improvement in reading abilities was found. From the results of EEG power spectra, no changes in the delta (1.5-3.5 Hz) frequency band were noted at any location. At CZ ($t=-2.65$; $p=0.03$) and P3 ($t=-2.51$; $p=0.04$), theta (4-7.5 Hz) increased. Paired sample correlations were significant and substantial (both $r=0.89$; $p=0.00$). This suggested that the increase was found in all participants. Alpha1 (8-11 Hz) was increased at CP4 ($t=-3.42$; $p=0.046$) and at O1 ($t=-2.55$; $p=0.04$). Paired sample correlations were $r=0.76$ ($p=0.03$) and $r=0.94$ ($p=0.00$), again suggesting an increase in all participants. Alpha1 (8-11 Hz) was intended to go down at locations F3 and T6. Beta1 (14.5-20 Hz) decreased at OZ ($t=-2.51$; $p=0.04$) with a paired sample correlation of $r=0.76$ ($p=0.03$). From the results of coherence training, in the alpha (8-13Hz) band, a consistent change was found at CP3/F3 ($t=-6.95$; $p=0.00$) with a paired sample correlation of $r=0.93$ ($p=0.00$). Other differences occurred at P3/F3, O1/F3 and C4/T4, showed insignificant correlations. A multitude of coherences went down in the beta (14.5-30 Hz) range, only P4/T4 ($t=3.62$; $p=0.01$) showed a consistent decrease among all subjects ($r=0.72$; $p=0.04$).

In summary, most studies reviewed in this section showed that neurofeedback could improve the learning abilities of children with dyslexia to certain extent, and also could modify those children's brain wave behaviors. However, the sample sizes of those studies were small. Therefore, it is necessary to increase the scale of similar studies in the future. Furthermore, different treatment protocols were used by different organizations. There is no unique protocol that can be proved to be effective in all cases to improve the learning abilities of all dyslexics or learning disabled children. It may be due to the fact that the causes of dyslexia are too complex so that we need to further sub-categorize the problem into finer and more specific cases. Because of this complexity, the inclusion/exclusion criteria for selecting subjects must be made with great care. It must exclude those subjects who have other neurological or psychiatric disorders such as ADHD. Otherwise, the result of a study may come from the result of curing of ADHD instead of curing of "developmental dyslexia" that is defined in this chapter.

Finally, the use of active task interventions in neurofeedback treatment was rare but it is worth investigating its efficacy in the future. In studies of [61] and [133], they showed that the performance of reading memory and auditory memory were greatly improved after neurofeedback treatment with active tasks intervention for those subjects who had traumatic brain injury, although those studies were clinical case studies only.

CONCLUSION

In this chapter, the principles and the process of neurofeedback treatment have been presented. The importance of a normative EEG/QEEG database and how it limits the EEG capturing methods, analysis methods, and ethnics and age of subjects in a neurofeedback treatment have also been discussed.

Then the considerations in selecting a normative EEG/QEEG database have been listed as initial guidelines for selecting a suitable database for starting a neurofeedback treatment or study. The considerations in selecting a normative database can be categorized into three main parts: technical considerations, analysis tools availability, and statistic information. According to these categories, four currently available normative EEG/QEEG databases are compared and their features are summarized in Table 1, Table 2, and Table 3.

And it seems that none of the four databases can be used for all circumstances. The selection mostly depends on what a user needs, for example active tasks, language included in reading and listening tasks, inverse calculation, ethnics of the targeting trainees, etc. Moreover, several concerns have been raised while discussing the considerations of selecting a normative database.

Firstly, in order to improve the spatial resolution of an inverse calculation, e.g. LORETA, it is necessary to increase the number of EEG electrodes. Secondly, a reliable and automatic artifact removal technique is required especially when active tasks intervention is involved during neurofeedback treatment.

Thirdly, the design of active task paradigms for dyslexics and efficacy of different paradigms have not been investigated. Fourthly, if reading and listening tasks are involved, the normative database becomes language and ethnic specific. This means that a separate database is required for a particular language. Fifthly, how the maximum processing latency from capturing EEG signal to the result fed back to the trainee that affects the efficiency of neurofeedback training has not been studied. Sixthly, there is no common agreement on how to handle the time-of-day factors while capturing EEG patterns for generating a normative database.

Finally, there is no common agreement on how to define the range of ages for each age group and the sample size of each age group so that the EEG/QEEG features within an age group are similar and statistical normal.

Reviewing neurofeedback for dyslexia is the second objective of this chapter. Eight studies have been reviewed including those subjects who were diagnosed with "learning disorder" that included reading, mathematics, and/or written expression disorders. After reviewing those studies, several points have been noticed. Firstly, although most studies showed that the learning abilities of those subjects have been improved to certain degree, no unique protocol can be proved to be effective to improve learning abilities for all cases. It may be due to the causes of dyslexia are complex.

Therefore, a further classification of learning disability is necessary. Secondly, the inclusion/exclusion criteria for selecting suitable subjects to be studied require a precise design. Thirdly, the efficacy and efficiency of using active tasks intervention in neurofeedback treatment for learning disabled children need to be studied in the future. Finally, more studies with larger scale should be done in the future in order to investigate the effectiveness of neurofeedback for dyslexia.

ACKNOWLEDGMENTS

This chapter was supervised and organized by Prof. Mary Waye (The Chinese University of Hong Kong, HK). I would like to thank Dr. Beverley Steffert (London University, UK) for her information about dyslexia and neurofeedback process. I would like to acknowledge Dr. David Kaiser (Sterman-Kaiser Imaging Laboratory) and Dr. Bernhard Wandernoth (HBImed AG) for their contributions and replies to the questionnaire. I would also like to thank Prof. Juri Kropotov (HBImed AG) for his information about QEEG analysis, and Ms. Alicia Chan for her information about statistics as well as proofreading this chapter.

APPENDIX

The original replies of the questionnaire from Sterman-Kaiser Imaging Laboratory and HBImed AG are attached in this appendix.

APPENDIX 1 – STERMAN-KAISER IMAGING LABORATORY

Questionnaire about Normative EEG/qEEG Databases

This is a questionnaire conducted by a Master of Science student from the Department of Biomedical Engineering in The Chinese University of Hong Kong. Now I am doing a study about neurofeedback for dyslexics. Please spend a couple of minutes to do the survey. The information provided will be published on some journals about neurotherapy if it is approved. Thank you.

Section 1. Background Information

1-1.	Name of your company/organization: Sterman-Kaiser Imaging Laboratory, Inc.
1-2.	Name of your normative EEG/qEEG database: SKIL 3 analysis and database

Section 2. Technical Specifications of Database

This section will ask you about the procedures of obtaining each subject's brain map.

2-1.	How many electrodes were used for capturing EEG raw data? 19 active, 2 linked ears, 1 ground
2-2.	What montage was used for capturing EEG raw data? Linked ears referential

2-3.	Does your database support re-montage after capturing raw EEG data? (Please specify what montage(s) can be achieved.) We have 4 montages that have databases: Referential (to linked ears), Hjorth Laplacian (local average, or 2-dimensional), Spherical Harmonic Expansion Laplacian (global average, or 3-dimensional/spheroid), and Brodmann area montage (a source or inverse solution). We also can display EEG in the following montages (but no spectral coefficient databases for these montages): Longitudinal bipolar, Transverse bipolar, Average, Average 17 (minus frontal poles), Vertex reference, Symmetry (8 homologue pairings), and Symphony, a montage I invented which assumes EEG of entire head or hemisphere is a single signal.
2-4.	What was the electrode impedance maintained while capturing raw EEG data? < 5K Ohm
2-5.	Does your database support amplifier matching procedure? (Please specify how to do it.) We will match to a new amplifier when given its specifications for new users. We will incorporate this into the code for all users from then on. Users can define frequency limits (e.g., only include 2-20 Hz or 1-30 Hz, etc, in coefficient calculation) and can alter site gains (weights) for both referential and laplacian montages.
2-6.	What was the sampling rate used while capturing raw EEG data? 1024 Hz down-sampled to 128 Hz, though we can analyze and export spectral coefficients up to 1024 Hz
2-7.	What was the bit resolution of those digitized raw EEG data? 16-bit
2-8.	What was the bandwidth of those captured raw EEG data? 1-38 Hz with single order filter at 1 Hz and much steeper filter at 38 Hz. For analysis we can analyze and export spectral coefficients from 0 to 64 Hz.
2-9.	What was the signal-to-noise ratio (SNR) of your EEG capturing system? (It should be tested by injecting a known sine wave into the input of the system.) – a known signal's amplitude is a bit of a fiction at the levels of electricity we are dealing with (microvolts), but my testing shows SNR of 12:1 to 18:1
2-10.	What EEG measuring system did you use for capturing raw EEG data? (Specify it if you used the one that is commercially available.) Lexicor and Neuronavigator

2-11.	What artifacts were removed or minimized after capturing raw EEG data? Muscle (including ocular) and equipment artifacts were removed using clipping technique where spectral coefficients of interval with artifacts were not included, with 75% overlapping windows
2-12.	What techniques or procedures were used for removing artifacts stated in Question 2-11? (For example, by inspection, using principal component analysis (PCA), using independent component analysis, etc.) Inspection with clipping technique
2-13.	Did the artifact removal procedures stated in Question 2-12 affect the characteristics of the original EEG signals, for example, their coherence and phase relationship? It would, but who knows as we did not analyze the unartifacted signal

Section 3. Analysis Tools

This section will ask you what analysis tools are included in your EEG/qEEG database package.

3-1.	Does it include eye closed or/and eye opened conditions? EC, EO baselines plus Math, and Reading tasks
3-2.	Does it include reading tasks? (Please specify what language(s) is/are available.) Yes, English, age-appropriate
3-3.	Please specify other language learning tasks if it includes. (For example, visual words recognition, auditory words recognition, memory-related tasks, etc.) none
3-4.	What time series analysis are included, waves, ERP, ERD/ERS? Spectral coefficient means
3-5.	Does it include spectral analysis both in one Hz resolution (single bin) and conventional brain wave rhythms (alpha, beta, theta, delta, and gamma)? (Please specify what are included.) Yes, 9 clinical or conventional bands (delta, theta, alpha, SMR (12-15 Hz), beta1-3, gamma 1-2, along with the integer Hz bands
3-6.	Does it support both relative and absolute measurements of voltage amplitude? Yes, along with entropy
3-7.	Does it support both relative and absolute measurements of spectral power? Spectral power output but no database for power, which is a measure of outlierness and a poor central tendency measure

3-8.	Does it support calculating the differences of voltage amplitudes between two arbitrary electrodes? yes
3-9.	Does it support calculating the ratios of two brain wave frequencies? (For example, alpha/beta, theta/alpha, theta/beta, etc.) Yes, and includes database of band magnitude ratios
3-10.	Does it support coherence and phase analysis? Yes, along with comodulation and magnitude unity (absolute asymmetry), the complements in the magnitude information of coherence and phase lag, respectively
3-11.	Does it contain some inverse problem calculations, for example, LORETA, sLORETA, ST-MAP, etc.? (Please specify the name(s) of the algorithm(s) that is/are supported.) Yes, a Brodmann area solution using Pasqual-Marqui's SHE Laplacian solution for the surface, then a proprietary inverse solution based on Brodmann area center points in Talaraich space with signals weighted according to Brodmann area size in square mm
3-12.	If the answer of Question 3-11 is yes, can it be calculated while capturing raw EEG data? (i.e. calculation in real-time) yes
3-13.	What is the maximum signal processing latency from sensor input to result output? (For example, from sensor to alpha wave, from sensor to spectrum, from sensor to coherence between two electrodes, from sensor to LORETA topology, etc.) This is an analysis program of a "dead" file – collected and closed file. For operant conditioning (real time) we have a ½-1/3rd second delay.

Section 4. Statistic Information

This section will ask you about the statistics of your EEG/qEEG database.

4-1.	What are the inclusion/exclusion criteria? (For example, ethnic, cultural backgrounds, geographic location, neurological and neuropsychological functioning, measures of intelligence, educational levels, gender, handedness, medication, etc.) Database is high educational levels for adults, 25% female, 75% male, no psychoactive medication, no history of neurological insult or psychiatric diagnosis, and normal sleep the previous day
4-2.	What is the age distribution? Children (6-11y), teens(12-19y), adults (20-75y), with future databases including tweens (10-13 y) and young adults (20-30 y)
4-3.	Is there any study or test of its sensitivity, specificity, reliability, Gaussian validation of different EEG parameters and clinical correlation? There is test-retest capability within the software.

4-4.	How to control the time-of-day factors? (For example, the time of measurement take place, before or after meals, etc.) We have time-of-day databases in previous versions of software, 21 one-hour intervals every 30 minutes from 8 am to 5 pm

The End of Questionnaire

Thanks for your contribution

You failed to ask about tapering window characteristics: we use Blackman-Harris -92db

You failed to ask whether we normalized the coherence distributions using a Fisher-z transformation, which we did for coherence, along with comodulation, and unity

We have automated artifact detection using Chi-square of voltage change distributions, which is quite effective.

Export parameters are the following:

Activity measures: magnitude, relative magnitude, magnitude variability, natural log magnitude, spectral entropy, automodulation, autocoherence, autophase, bimodulation, band ratio, bicoherence, biphase,and two measures I invented rogue site analysis and rogue phase analysis; Connectivity measures: coherence, phase lag, comodulation, unity (absolute asymmetry), and corticality (comodulation-coherence in z-scores) – see http://skiltopo.com /period.htm

We also have user-input graphical mapping capability, so anyone can input values and generate activity, connectivity, and Brodmann area maps from their own data or summary data.

Graphically we have 5 presentations: numerical tables, spectral plots, topometrics (frequency line graphs), topographs (brain maps), and network maps (brain maps showing 19x19 comparisons comparisons).

We provide trending capability for magnitude, entropy, site-centered coherence, site-centered comodulation for referential, Hjorth and Laplacian montages.

We also have clinical report generation with a proprietary but open Brodmann area function database.

APPENDIX 2 – HBIMED AG

Questionnaire about Normative EEG/qEEG Databases

This is a questionnaire conducted by a Master of Science student from the Department of Biomedical Engineering in The Chinese University of Hong Kong. Now I am doing a study about neurofeedback for dyslexics. Please spend a couple of minutes to do the survey. The information provided will be published on some journals about neurotherapy if it is approved. Thank you.

Section 1. Background Information

1-1.	Name of your company/organization: HBImed AG Untere Gasse 17 7000 Chur Switzerland info@hbimed.com www.hbimed.com
1-2.	Name of your normative EEG/qEEG database: HBI reference database

Section 2. Technical Specifications of Database

This section will ask you about the procedures of obtaining each subject's brain map.

2-1.	How many electrodes were used for capturing EEG raw data? 19
2-2.	What montage was used for capturing EEG raw data? Linked ears.
2-3.	Does your database support re-montage after capturing raw EEG data? (Please specify what montage(s) can be achieved.) Yes. Three montages: linked ears, common reference, local reference (Hjorth).
2-4.	What was the electrode impedance maintained while capturing raw EEG data? Below 5 kOhm
2-5.	Does your database support amplifier matching procedure? (Please specify how to do it.) The soft reads many data formats, including EDF+
2-6.	What was the sampling rate used while capturing raw EEG data? 250 or 500 Samples per second per channel
2-7.	What was the bit resolution of those digitized raw EEG data? 16 bit ADC
2-8.	What was the bandwidth of those captured raw EEG data? 0.16 – 70 Hz

2-9.	What was the signal-to-noise ratio (SNR) of your EEG capturing system? (It should be tested by injecting a known sine wave into the input of the system.) <1.5 mcV
2-10.	What EEG measuring system did you use for capturing raw EEG data? (Specify it if you used the one that is commercially available.) Mitsar-EEG-201. Is commercially available by the Mitsar, Ltd, Saint Petersburg, Russia
2-11.	What artifacts were removed or minimized after capturing raw EEG data? Eyes blink artifacts were corrected by ICA. Slow and fast waves were automatically excluded.
2-12.	What techniques or procedures were used for removing artifacts stated in Question 2-11? (For example, by inspection, using principal component analysis (PCA), using independent component analysis, etc.) Using ICA, thresholds for fast and slow waves.
2-13.	Did the artifact removal procedures stated in Question 2-12 affect the characteristics of the original EEG signals, for example, their coherence and phase relationship? Three methods of artifact removal were compared and the results show that only mild distortion is observed.

Section 3. Analysis Tools

This section will ask you what analysis tools are included in your EEG/qEEG database package.

3-1.	Does it include eye closed or/and eye opened conditions? Yes
3-2.	Does it include reading tasks? (Please specify what language(s) is/are available.) Yes, in German.
3-3.	Please specify other language learning tasks if it includes. (For example, visual words recognition, auditory words recognition, memory-related tasks, etc.) Altogether 6 task conditions: GO/NOGO, emotion face recognition, reading task, auditory, three stimulus odd ball + novelty auditory task
3-4.	What time series analysis are included, waves, ERP, ERD/ERS? ERP, ERD/ERS, wavelet
3-5.	Does it include spectral analysis both in one Hz resolution (single bin) and conventional brain wave rhythms (alpha, beta, theta, delta, and gamma)? (Please specify what are included.) Yes, both conventional EEG bands and 1 Hz bands.

3-6.	Does it support both relative and absolute measurements of voltage amplitude? Yes.
3-7.	Does it support both relative and absolute measurements of spectral power? Yes
3-8.	Does it support calculating the differences of voltage amplitudes between two arbitrary electrodes? Yes
3-9.	Does it support calculating the ratios of two brain wave frequencies? (For example, alpha/beta, theta/alpha, theta/beta, etc.) Yes
3-10.	Does it support coherence and phase analysis? Yes
3-11.	Does it contain some inverse problem calculations, for example, LORETA, sLORETA, ST-MAP, etc.? (Please specify the name(s) of the algorithm(s) that is/are supported.) Yes, LORETA and sLORETA, dipole approximation
3-12.	If the answer of Question 3-11 is yes, can it be calculated while capturing raw EEG data? (i.e. calculation in real-time) No, but it can be easily Three montages: linked ears, common reference, local reference (Hjorth) done.
3-13.	What is the maximum signal processing latency from sensor input to result output? (For example, from sensor to alpha wave, from sensor to spectrum, from sensor to coherence between two electrodes, from sensor to LORETA topology, etc.) Almost on-line.

Section 4. Statistic Information

This section will ask you about the statistics of your EEG/qEEG database.

4-1.	What are the inclusion/exclusion criteria? (For example, ethnic, cultural backgrounds, geographic location, neurological and neuropsychological functioning, measures of intelligence, educational levels, gender, handedness, medication, etc.) No history of seizers, No psychiatric or neurological disease as well as no history of such disease in family, average of higher marks in school, successful performance at work.
4-2.	What is the age distribution? From 7 to 89 years. Children 7-17 – 300 subjects. Adults 18-60 – 500. Elderly >60 - 200
4-3.	Is there any study or test of its sensitivity, specificity, reliability, Gaussian validation of different EEG parameters and clinical correlation? Not with our database. There is plenty of literature for others.

4-4.	How to control the time-of-day factors? (For example, the time of measurement take place, before or after meals, etc.) No control.

The End of Questionnaire.

Thanks for your contribution.

REFERENCES

[1] World Health Organization. ICD-10 Version: 2010. [Online]. 2012. Available from: http://apps.who.int/classifications/icd10/browse/2010/en#/F80-F89.

[2] Walker, JE; Norman, CA. The neurophysiology of dyslexia: a selective review with implications for neurofeedback remediation and results of treatment in twelve consecutive patients. *J. Neurother.*. 2006 Dec; 10(1): p. 45-55.

[3] Vernon, D; Frick, A; Gruzelier, J. Neurofeedback as a treatment for ADHD: a methodological review with implications for future research. *J. Neurother*. 2004 May; 8(2): p. 53-82.

[4] PsychiatryOnline 315.00 Reading Disorder. [Online]. 2012. Available from: http://psycho.silverchair.com/content.aspx?aID=7313andsearchStr=dyslexia.

[5] Démonet, JF; Taylor, MJ; Chaix, Y. Developmental dyslexia. *Lancet*. 2004 May; 363(9419): p. 1451-1460.

[6] Richlan, F; Kronbichler, M; Wimmer, H. Structural abnormalities in the dyslexic brain: A meta-analysis of voxel-based morphometry studies. *Hum. Brain Mapp*. 2012 Jun; (Epub ahead of print).

[7] Vandermosten, M; Boets, B; Wouters, J; Ghesquière, P. A qualitative and quantitative review of diffusion tensor imaging studies in reading and dyslexia. *Neurosci. Biobehav. Rev*. 2012 Jul; 36(6): p. 1532-1552.

[8] Caylak, E. Neurobiological approaches on brains of children with dyslexia: review. *Acad. Radiol*. 2009 Aug, 16(8): p. 1003-1024.

[9] Habib, M. The neurological basis of developmental dyslexia: an overview and working hypothesis. *Brain*. 2000 Dec; 123: p. 2373-2399.

[10] Eckert, MA; Leonard, CM; Wilke, M; Eckert, M; Richards, T; Richards, A; et al. Anatomical signatures of dyslexia in children: unique information from manual and voxel based morphometry brain measures. *Cortex*. 2005 Jun; 41(3): p. 304-315.

[11] Vinckenbosch, E; Robichon, F; Eliez, S. Gray matter alteration in dyslexia: converging evidence from volumetric and voxel-by-voxel MRI analyses. *Neuropsychologia*. 2005; 43(3): p. 324-331.

[12] Hoeft, F; Meyler, A; Hernandez, A; Juel, C; Taylor-Hill, H; Martindale, JL; et al. Functional and morphometric brain dissociation between dyslexia and reading ability. *Proc. Natl. Acad. Sci. USA*. 2007 Mar; 104(10): p. 4234-4239.

[13] Jäncke, L; Siegenthaler, T; Preis, S; Steinmetz, H. Decreased white-matter density in a left-sided fronto-temporal network in children with developmental language disorder: evidence for anatomical anomalies in a motor-language network. *Brain Lang*. 2007 Jul; 102(1): p. 91-98.

[14] Badcock, NA; Bishop, DV; Hardiman, MJ; Barry, JG; Watkins, KE. Co-localisation of abnormal brain structure and function in specific language impairment. *Brain Lang.* 2012 Mar; 120(3): p. 310-320.

[15] Hasan, KM; Molfese, DL; Walimuni, IS; Stuebing, KK; Papanicolaou, AC; Narayana, PA; et al. Diffusion tensor quantification and cognitive correlates of the macrostructure and microstructure of the corpus callosum in typically developing and dyslexic children. *NMR Biomed.* 2012 Mar; (Epub ahead of print).

[16] Raschle, NM; Zuk, J; Gaab, N. Functional characteristics of developmental dyslexia in left-hemispheric posterior brain regions predate reading onset. *Proc. Natl. Acad. Sci. USA*. 2012 Feb; 109(6): p. 2156-2161.

[17] Kovelman, I; Norton, ES; Christodoulou, JA; Gaab, N; Lieberman, DA; Triantafyllou, C; et al. Brain basis of phonological awareness for spoken language in children and its disruption in dyslexia. *Cereb Cortex*. 2012 Apr; 22(4): p. 754-764.

[18] Richlan, F; Kronbichler, M; Wimmer, H. Meta-analyzing brain dysfunctions in dyslexic children and adults. *Neuroimage*. 2011 Jun; 56(3): p. 1735-1742.

[19] Kropotov, JD. *Quantitative EEG, Event-Related Potentials and Neurotherapy* U.S.: Academic Press; 2009.

[20] Thatcher, RW; Lubar, JF. History of the scientific standards of QEEG normative databases. In: Budzynski TH, Budzynski HK, Evans JR, Abarbanel A, editors. *Introduction to Quantitative EEG and Neurofeedback: Advanced Theory and Applications*. 2nd ed. U.S.: Academic Press; 2008.

[21] Thornton, KE; Carmody, DP. Eyes-closed and activation QEEG databases in predicting cognitive effectiveness and the inefficiency hypothesis. *J. Neurother.* 2009 Jan; 13(1): p. 1-21.

[22] Key, AP; Dove, GO; Maguire, MJ. Linking brainwaves to the brain: an ERP primer. *Dev. Neuropsychol.* 2005; 27(2): p. 183-215.

[23] Näätänen, R; Gaillard, AW; Mäntysalo, S. Early selective-attention effect on evoked potential reinterpreted. *Acta. Psychol. (Amst).* 1978 Jul; 42(4): p. 313-329.

[24] Khan, A; Hämäläinen, JA; Leppänen, PH; Lyytinen, H. Auditory event-related potentials show altered hemispheric responses in dyslexia. *Neurosci. Lett.* 2011 Jul; 498(2): p. 127-132.

[25] Neuhoff, N; Bruder, J; Bartling, J; Warnke, A; Remschmidt, H; Müller-Myhsok, B; et al. Evidence for the late MMN as a neurophysiological endophenotype for dyslexia. *PLoS One*. 2012; 7(5): p. e34909.

[26] Dhar, M; Been, PH; Minderaa, RB; Althaus, M. Reduced interhemispheric coherence in dyslexic adults. *Cortex*. 2010 Jun; 46(6): p. 794-798.

[27] Ramus, F; Rosen, S; Dakin, SC; Day, BL; Castellote, JM; White, S; et al. Theories of developmental dyslexia: insights from a multiple case study of dyslexic adults. *Brain*. 2003 Apr; 126: p. 841-865.

[28] Tallal, P. Auditory temporal perception, phonics, and reading disabilities in children. *Brain Lang*. 1980 Mar; 9(2): p. 182-198.

[29] Tallal, P; Miller, S; Fitch, RH. Neurobiological basis of speech: a case for the preeminence of temporal processing. *Ann N Y Acad Sci*. 1993 Jun; 682: p. 27-47.

[30] McAnally, KI; Stein, JF. Auditory temporal coding in dyslexia. *Proc. Biol. Sci.* 1996 Aug; 263(1373): p. 961-965.

[31] Ahissar, M; Protopapas, A; Reid, M; Merzenich, MM. Auditory processing parallels reading abilities in adults. *Proc Natl Acad Sci USA*. 2000 Jun; 97(12): p. 6832-6837.

[32] Nagarajan, S; Mahncke, H; Salz, T; Tallal, P; Roberts, T; Merzenich, MM. Cortical auditory signal processing in poor readers. *Proc. Natl. Acad. Sci. USA*. 1999 May; 96(11): p. 6483-6488.

[33] Stein, J; Walsh, V. To see but not to read; the magnocellular theory of dyslexia. *Trends Neurosci*. 1997 Apr; 20(4): p. 147-152.

[34] Hari, R; Renvall, H; Tanskanen, T. Left minineglect in dyslexic adults. *Brain*. 2001 Jul; 124: p. 1373-1380.

[35] Livingstone, MS; Rosen, GD; Drislane, FW; Galaburda, AM. Physiological and anatomical evidence for a magnocellular defect in developmental dyslexia. *Proc. Natl. Acad. Sci. USA*. 1991 Sep; 88(18): p. 7943-7947.

[36] Lovegrove, WJ; Bowling, A; Badcock, D; Blackwood, M. Specific reading disability: differences in contrast sensitivity as a function of spatial frequency. *Science*. 1980 Oct; 210(4468): p. 439-440.

[37] Cornelissen, P; Richardson, A; Mason, A; Fowler, S; Stein, J. Contrast sensitivity and coherent motion detection measured at photopic luminance levels in dyslexics and controls. *Vision Res*. 1995 May; 35(10): p. 1483-1494.

[38] Eden, GF; VanMeter, JW; Rumsey, JM; Maisog, JM; Woods, RP; Zeffiro, TA. Abnormal processing of visual motion in dyslexia revealed by functional brain imaging. *Nature*. 1996 Jul; 382(6586): p. 66-69.

[39] Nicolson, RI; Fawcett, AJ. Automaticity: a new framework for dyslexia research? *Cognition*. 1990 May; 35(2): p. 159-182.

[40] Nicolson, R; Fawcett, AJ; Dean, P. Dyslexia, development and the cerebellum. *Trends Neurosci*. 2001 Sep; 24(9): p. 515-516.

[41] Fawcett, AJ; Nicolson, RI. *The Dyslexia Screening Test* London: Psychological Corporation; 1996.

[42] Nicolson, RI; Fawcett, AJ; Dean, P. Time estimation deficits in developmental dyslexia: evidence of cerebellar involvement. *Proc. Biol. Sci*. 1995 Jan; 259(1354): p. 43-47.

[43] Arns, M; Peters, S; Breteler, R; Verhoeven, L. Different brain activation patterns in dyslexic children: evidence from EEG power and coherence patterns for the double-deficit theory of dyslexia. *J. Integr. Neurosci*. 2007 Mar; 6(1): p. 175-190.

[44] Siok, WT; Perfetti, CA; Jin, Z; Tan, LH. Biological abnormality of impaired reading is constrained by culture. *Nature*. 2004 Sep; 431(7004): p. 71-76.

[45] DeFrancis, J. *Visible Speech: The Diverse Oneness Of Writing Systems* Honolulu: University of Hawaii Press; 1989.

[46] Mattingly, IG. Reading, linguistic awareness, and language acquisition. In: Valtin JD, editor. *Language Awareness And Learning To Read*. New York: Springer; 1984.

[47] Chung, KK; McBride-Chang, C; Wong, SW; Cheung, H; Penney, TB; Ho, CS. The role of visual and auditory temporal processing for Chinese children with developmental dyslexia. *Ann of Dyslexia*. 2008 Jun; 58(1): p. 15-35.

[48] Perfetti, CA; Tan, LH. The constituency model of Chinese word identification. In: Wang J, Inhoff AW, Chen I, editors. *Reading Chinese Script: A Cognitive Analysis*. New Jersey: Lawrence Erlbaum Associates; 1999.

[49] Perfetti, CA; Liu, Y; Tan, LH. The lexical constituency model: some implications of research on Chinese for general theories of reading. *Psychol. Rev.* 2005 Jan; 112(1): p. 43-59.

[50] Ho, CS; Chan, DW; Lee, SH; Tsang, SM; Luan, VH. Cognitive profiling and preliminary subtyping in Chinese developmental dyslexia. *Cognition.* 2004 Feb; 91(1): p. 43-75.

[51] Ho, CS; Chan, DW; Tsang, SM; Lee, SH. The cognitive profile and multiple-deficit hypothesis in Chinese developmental dyslexia. *Dev. Psychol.* 2002 Jul; 38(4): p. 543-553.

[52] Courtney, SM; Petit, L; Maisog, JM; Ungerleider, LG; Haxby, JV. An area specialized for spatial working memory in human frontal cortex. *Science.* 1998 Feb; 279(5355): p. 1347-1351.

[53] D'Esposito, M; Detre, JA; Alsop, DC; Shin, RK; Atlas, S; Grossman, M. The neural basis of the central executive system of working memory. *Nature.* 1995 Nov; 378(6554): p. 279-281.

[54] Petrides, M; Alivisatos, B; Meyer, E; Evans, AC. Functional activation of the human frontal cortex during the performance of verbal working memory tasks. *Proc. Natl. Acad. Sci. USA.* 1993; 90: p. 878-882.

[55] Siok, WT; Niu, Z; Jin, Z; Perfetti, CA; Tan, LH. A structural-functional basis for dyslexia in the cortex of Chinese readers. *Proc. Natl. Acad. Sci. USA.* 2008 Apr; 105(14): p. 5561-5566.

[56] Chung, KK; Tong, X; McBride-Chang, C. Evidence for a deficit in orthographic structure processing in Chinese character recognition among developmental dyslexics: An Event-Related Potentials (ERP) study. *Brain Res.* 2012 Jun; (Epub ahead of print).

[57] Demos, JN. *Getting Started with Neurofeedback* New York: W. W. Norton and Company, Inc.; 2005.

[58] Thornton, KE; Carmody, DP. Electroencephalogram biofeedback for reading disability and traumatic brain injury. *Child Adolesc. Psychiatr. Clin. N Am.* 2005 Jan; 14(1): p. 137-162.

[59] Egner, T; Gruzelier, JH. Learned self-regulation of EEG frequency components affects attention and event-related brain potentials in humans. *Neuroreport.* 2001 Dec; 12(18): p. 4155-4159.

[60] Vernon, D; Egner, T; Cooper, N; Compton, T; Neilands, C; Sheri, A; et al. The effect of training distinct neurofeedback protocols on aspects of cognitive performance. *Int. J. Psychophysiol.* 2003 Jan; 47(1): p. 75-85.

[61] Thornton, K. Improvement/rehabilitation of memory functioning with neurotherapy /QEEG biofeedback. *J. Head Trauma Rehabil.* 2000 Dec; 15(6): p. 1285-1296.

[62] Tinius, TP; Tinius, KA. Changes after EEG biofeedback and cognitive retraining in adults with mild traumatic brain injury and attention deficit hyperactivity disorder. *J. Neurother.* 2000 Jun; 4(2): p. 27-44.

[63] Sterman, MB; Macdonald, LR. Effects of central cortical EEG feedback training on incidence of poorly controlled seizures. *Epilepsia.* 1978 Jun; 19(3): p. 207-222.

[64] Sterman, MB; Macdonald, LR; Stone, RK. Biofeedback training of the sensorimotor electroencephalogram rhythm in man: effects on epilepsy. *Epilepsia.* 1974 Sep; 15(3): p. 395-416.

[65] Uhlmann, C; Fröscher, W. Biofeedback treatment in patients with refractory epilepsy: changes in depression and control orientation. *Seizure*. 2001 Jan; 10(1): p. 34-38.

[66] Gruzelier, J; Hardman, E; Wild, J; Zaman, R. Learned control of slow potential interhemispheric asymmetry in schizophrenia. *Int. J. Psychophysiol.* 1999 Dec; 34(3): p. 341-348.

[67] Sherlin, LH. Diagnosing and treating brain function through the use of low resolution brain electromagnetic tomography (LORETA). In: Budzynski TH, Budzynski HK, Evans JR, Abarbanel A, editors. *Introduction to Quantitative EEG and Neurofeedback: Advanced Theory and Applications*. 2nd ed. U.S.: Academic Press; 2008.

[68] Kandel, ER. Nerve cells and behavior. In: Kandel ER, Schwartz JH, Jessell TM, editors. *Principles of Neural Science*. 4th ed. New York: McGraw-Hill; 2000.

[69] Nunez, PL; Srinivasan, R. *Electric Fields of the Brain: The Neurophysics of EEG*. 2nd ed. New York: Oxford University Press; 2006.

[70] Pascual-Marqui, RD; Michel, CM; Lehmann, D. Low resolution electromagnetic tomography: a new method for localizing electrical activity in the brain. *Int. J. Psychophysiol.* 1994 Oct; 18(1): p. 49-65.

[71] Johnstone, J; Gunkelman, J. Use of databases in QEEG evaluation. In: Lubar JF, editor. *Quantitative Electroencephalographic Analysis (QEEG) Databases for Neurotherapy: Description, Validation, and Application.* New York: The Haworth Medical Press; 2003.

[72] Thatcher, RW; Biver, CJ; North, D. EEG coherence and phase delays: comparisons between single reference, average reference and current source density. [Online]. 2004. Available from: http://www.appliedneuroscience.com/COMPARISONS-COMMONREF-AVE-LAPLACIAN.pdf.

[73] Thatcher, RW. Normative EEG databases and EEG biofeedback. *J. Neurother.* 1998 Mar; 2(4): p. 8-39.

[74] EEGLAB - Open Source Matlab Toolbox for Psychophysiological Research. [Online]. 2010. Available from: http://sccn.ucsd.edu/~scott/ica.html.

[75] Ifeachor, EC; Jervis, BW. *Digital Signal Processing: A Practical Approach* U.S.: Addison-Wesley Publishing Company; 1993.

[76] Nunez, P. *Electrical Fields of the Brain* Cambridge: Oxford Univ Press; 1981.

[77] Nunez, P. *Neocortical Dynamics and Human EEG Rhythms* New York: Oxford Univ Press; 1995.

[78] Coghill, D. Neurofeedback training improves ADHD symptoms more than attention skills training. *Evid Based Ment Health*. 2010 Feb; 13(1): p. 21.

[79] Fernández, T; Herrera, W; Harmony, T; Díaz-Comas, L; Santiago, E; Sánchez, L; et al. EEG and behavioral changes following neurofeedback treatment in learning disabled children. *Clin Electroencephalogr*. 2003 Jul; 34(3): p. 145-152.

[80] Fernández, T; Harmony, T; Fernández-Bouzas, A; Díaz-Comas, L; Prado-Alcalá, RA; Valdés-Sosa, P; et al. Changes in EEG current sources induced by neurofeedback in learning disabled children. An exploratory study. *Appl. Psychophysiol. Biofeedback*. 2007 Dec; 32(3-4): p. 169-183.

[81] Schummer, GJ. The disconnection syndrome. *Biofeedback*. 2008; 36(4): p. 157-162.

[82] Grech, R; Cassar, T; Muscat, J; Camilleri, KP; Fabri, SG; Zervakis, M; et al. Review on solving the inverse problem in EEG source analysis. *J. Neuroeng. Rehabil.* 2008 Nov; 5(25).

[83] Pascual-Marqui, RD. Discrete, 3D distributed linear imaging methods of electric neuronal activity. Part 1: exact, zero error localization. [Online]. 2007. Available from: http://www.uzh.ch/keyinst/eLORETA/eLor0errLikeItOrNot.pdf.

[84] Pascual-Marqui, RD. Standardized low resolution brain electromagnetic tomography (sLORETA): technical details. [Online]. 2002. Available from: http://www.uzh.ch/keyinst/NewLORETA/sLORETA/sLORETA-Math01.pdf.

[85] Thatcher, RW; North, D; Biver, C. Parametric vs. non-parametric statistics of low resolution electromagnetic tomography (LORETA). *Clin EEG Neurosci.* 2005 Jan; 36(1): p. 1-8.

[86] Thatcher, RW; North, D; Biver, C. Evaluation and validity of a LORETA normative EEG database. *Clin EEG Neurosci.* 2005 Apr; 36(2): p. 116-122.

[87] Huizenga, HM; de Munck, JC; Waldorp, LJ; Grasman, RP. Spatiotemporal EEG/MEG source analysis based on a parametric noise covariance model. *IEEE Trans Biomed Eng.* 2002 Jun; 49(6): p. 533-539.

[88] Hori, J; He, B. Equivalent dipole source imaging of brain electric activity by means of parametric projection filter. *Ann. Biomed. Eng.* 2001 May; 29(5): p. 436-445.

[89] Waldorp, LJ; Huizenga, HM; Dolan, CV; Molenaar, PC. Estimated generalized least squares electromagnetic source analysis based on a parametric noise covariance model. *IEEE Trans Biomed. Eng.* 2001 Jun; 48(6): p. 737-741.

[90] Bosch-Bayard, J; Valdés-Sosa, P; Virues-Alba, T; Aubert-Vázquez, E; John, ER; Harmony, T; et al. 3D statistical parametric mapping of EEG source spectra by means of variable resolution electromagnetic tomography (VARETA). *Clin. Electroencephalogr.* 2001 Apr; 32(2): p. 47-61.

[91] Machado, C; Cuspineda, E; Valdés, P; Virues, T; Llopis, F; Bosch, J; et al. Assessing acute middle cerebral artery ischemic stroke by quantitative electric tomography. *Clin EEG Neurosci.* 2004 Jul; 35(3): p. 116-124.

[92] Congedo, M; Lubar, JF; Joffe, D. Low-resolution electromagnetic tomography neurofeedback. *IEEE Trans Neural Syst Rehabil Eng.* 2004 Dec; 12(4): p. 387-397.

[93] Cannon, R; Lubar, J; Congedo, M; Thornton, K; Towler, K; Hutchens, T. The effects of neurofeedback training in the cognitive division of the anterior cingulate gyrus. *Int. J. Neurosci.* 2007 Mar; 117(3): p. 337-357.

[94] Cannon, R; Lubar, J; Sokhadze, E; Baldwin, D. LORETA neurofeedback for addiction and the possible neurophysiology of psychological processes influenced: a case study and region of interest analysis of LORETA neurofeedback in right anterior cingulate cortex. *J. Neurother.* 2008 Dec; 12(4): p. 227-241.

[95] Hogg, RV; Ledolter, J. *Applied Statistics for Engineers and Physical Scientists.* 2nd ed. New York: Macmillan Publishing Company; 1992.

[96] Congedo, M; Lubar, JF. Parametric and non-parametric analysis of QEEG: normative database comparisons in electroencephalography, a simulation study on accuracy. In: Lubar JF, editor. *Quantitative Electroencephalographic Analysis (QEEG) Databases for Neurotherapy: Description, Validation, and Application.* New York: The Haworth Medical Press; 2003.

[97] Thatcher, RW; Walker, RA; Biver, CJ; North, DN; Curtin, R. Quantitative EEG normative databases: validation and clinical correlation. In: Lubar JF, editor. *Quantitative Electroencephalographic Analysis (QEEG) Databases for Neurotherapy:*

Description, Validation, and Application. New York: The Haworth Medical Press; 2003.

[98] Lorensen, TD; Dickson, P. Quantitative EEG normative databases: a comparative investigation. In: Lubar JF, editor. *Quantitative Electroencephalographic Analysis (QEEG) Databases for Neurotherapy: Description, Validation, and Application*. New York: The Haworth Medical Press; 2003.

[99] BRAINnet - Brain Research and Integrative Neuroscience Network. [Online]. 2010. Available from: http://www.brainnet.net/about/brain-resource-international-database/.

[100] Brain Resource Ltd. Personalized Medicine - Depression Treatment - Brain Database. [Online]. 2010. Available from: http://www.brainresource.com/about_us/index.php?id=25.

[101] NeuroGuide EEG and QEEG Software, Applied Neuroscience, Inc. [Online]. 2010. Available from: http://www.appliedneuroscience.com/NG-BrochureWEB.htm.

[102] NeuroGuide EEG and QEEG Software, Applied Neuroscience, Inc. NeuroGuide Manual. [Online]. 2008. Available from: http://www.appliedneuroscience.com/NeuroGuide_Deluxe.pdf.

[103] Thatcher, RW; Walker, RA; Biver, CJ; North, DM; Curtin, R. Sensitivity and specificity of an EEG normative database: validation and clinical correlation. [Online]. 2010. Available from: http://www.appliedneuroscience.com/LSNDBweb.pdf.

[104] Thatcher, RW. Nineteen channel quantitative EEG and EEG biofeedback. [Online]. 2010. Available from: http://www.appliedneuroscience.com/Biofeedback-19Channels.pdf.

[105] Gordon, E; Cooper, N; Rennie, C; Hermens, D; Williams, LM. Integrative neuroscience: the role of a standardized database. *Clin EEG Neurosci*. 2005 Apr; 36(2): p. 64-75.

[106] Williams, LM; Simms, E; Clark, CR; Paul, RH; Rowe, D; Gordon, E. The test-retest reliability of a standardized neurocognitive and neurophysiological test battery: "neuromarker". *Int. J. Neurosci*. 2005 Dec; 115(12): p. 1605-1630.

[107] Paul, RH; Gunstad, J; Cooper, N; Williams, LM; Clark, CR; Cohen, RA; et al. Cross-cultural assessment of neuropsychological performance and electrical brain function measures: additional validation of an international brain database. *Int. J. Neurosci*. 2007 Apr; 117(4): p. 549-568.

[108] Gratton, G; Coles, MG; Donchin, E. A new method for off-line removal of ocular artifact. *Electroencephalogr Clin Neurophysiol*. 1983 Apr; 55(4): p. 468-484.

[109] Tansey, MA. Wechsler (WISC-R) changes following treatment of learning disabilities via EEG biofeedback training in a private practice setting. *Aust. J. Psychol*. 1991; 43: p. 147-153.

[110] Tansey, MA. Righting the rhythms of reason: EEG biofeedback training as a therapeutic modality in a clinical office setting. *Medical Psychotherapy*. 1990; 3: p. 57-68.

[111] Lubar, JF; Bianchini, KJ; Calhoun, WH; Lambert, EW; Brody, ZH; Shabsin, HS. Spectral analysis of EEG differences between children with and without learning disabilities. *J Learn Disabil*. 1985 Aug-Sep; 18(7): p. 403-408.

[112] Tansey, MA. EEG sensorimotor rhythm biofeedback training: some effects on the neurologic precursors of learning disabilities. *Int. J. Psychophysiol*. 1984 Feb; 1(2): p. 163-177.

[113] Tansey, MA. Brainwave signatures - an index reflective of the brain's functional neuroanatomy: further findings on the effect of EEG sensorimotor rhythm biofeedback training on the neurologic precursors of learning disabilities. *Int. J. Psychophysiol.* 1985 Nov; 3(2): p. 85-99.

[114] Kaufman, AS. *Intelligent Testing with the WISC-R* New York: Wiley; 1979.

[115] Wechsler, D. *Manual for the Wechsler Intelligence Scale for Children - Revised* New York: Psychological Corporation; 1974.

[116] Fenger, TN. Visual-motor integration and its relation to EEG neurofeedback brain wave patterns, reading, spelling, and arithmetic achievement in attention deficit disordered and learning disabled students. *J. Neurother.* 1998; 3(1): p. 9-18.

[117] Wikler, A; Dixon, JF; Parker, JB. Brain function in problem children and controls: psychometric, neurological, and electroencephalographic comparisons. *Am. J. Psychiatry.* 1970 Nov; 127(5): p. 634-645.

[118] Mann, CA; Lubar, JF; Zimmerman, AW; Miller, CA; Muenchen, RA. Quantitative analysis of EEG in boys with attention-deficit-hyperactivity disorder: controlled study with clinical implications. *Pediatr. Neurol.* 1992 Jan-Feb; 8(1): p. 30-36.

[119] Satterfield, JH; Braley, BW. Evoked potentials and brain maturation in hyperactive and normal children. *Electroencephalogr Clin Neurophysiol.* 1977 Jul; 43(1): p. 43-51.

[120] Linden, M. Event related potentials of subgroups of attention deficit disorder children and implications for EEG biofeedback. *California Biofeedback.* 1991; 7(1): p. 7-12.

[121] La Vaque, TJ; Rossiter, T. The ethical use of placebo controls in clinical research: the Declaration of Helsinki. *Appl Psychophysiol Biofeedback.* 2001 Mar; 26(1): p. 23-37.

[122] Harmony, T; Marosi, E; Díaz de León, AE; Becker, J; Fernández, T. Effect of sex, psychosocial disadvantages and biological risk factors on EEG maturation. *Electroencephalogr Clin. Neurophysiol.* 1990 Jun; 75(6): p. 482-491.

[123] John, ER; Prichep, L; Ahn, H; Easton, P; Fridman, J; Kaye, H. Neurometric evaluation of cognitive dysfunctions and neurological disorders in children. *Prog. Neurobiol.* 1983; 21(4): p. 239-290.

[124] Alvarez, A; Pérez-Avalo, MC; Morenza, L. Neuropsychological assessment of learning-disorder children with paroxysmal EEG activity. In: Magistretti PJ, Leonardi M, editors. *New Issues in Neurosciences: Basic and Clinical Approaches. Neuropsychological Aspects in Epilepsy.* New York: Thieme; 1992.

[125] Fernández, T; Harmony, T; Silva, J; Galán, L; Díaz-Comas, L; Bosch, J; et al. Relationship of specific EEG frequencies at specific brain areas with performance. *Neuroreport.* 1998 Nov; 9(16): p. 3681-3687.

[126] Fernández, T; Harmony, T; Silva-Pereyra, J; Fernández-Bouzas, A; Gersenowies, J; Galán, L; et al. Specific EEG frequencies at specific brain areas and performance. *Neuroreport.* 2000 Aug; 11(12): p. 2663-2668.

[127] Becerra, J; Fernández, T; Harmony, T; Caballero, MI; García, F; Fernández-Bouzas, A; et al. Follow-up study of learning-disabled children treated with neurofeedback or placebo. *Clin. EEG Neurosci.* 2006 Jul; 37(3): p. 198-203.

[128] Breteler, MH; Arns, M; Peters, S; Giepmans, I; Verhoeven, L. Improvements in spelling after QEEG-based neurofeedback in dyslexia: a randomized controlled treatment study. *Appl. Psychophysiol. Biofeedback.* 2010 Mar; 35(1): p. 5-11.

[129] Wentink, H; Verhoeven, L. *Protocol Leesproblemen en Dyslexie* Nijmegen: Expertise Centrum Nederlands; 2003.

[130] Instituut voor Orthopedagogiek and Cito. *Normeringsonderzoek Screeningstest voor Taal- en Leesproblemen basisonderwijs jaargroep 1 tot en met 8* Nijmegen/Arnhem: Instituut voor orthopedagogiek/Cito; 2004.

[131] Geelhoed, J; Reitsma, P. *PI-dictee* Lisse: Swets and Zeitlinger; 1999.

[132] Breteler, MH; Arns, M; Peters, S; Giepmans, I; Verhoeven, L. Erratum to: Improvements in spelling after QEEG-based neurofeedback in dyslexia: a randomized controlled treatment study. *Appl. Psychophysiol. Biofeedback*. 2010 Jun; 35(2): p. 187.

[133] Thornton, KE. The improvement/rehabilitation of auditory memory functioning with EEG biofeedback. *NeuroRehabilitation*. 2002; 17(1): p. 69-80.

[134] Iglesias, A; Derman, B. *Prueba de lecto-escritura* México D. F.: Progreso; 1985.

[135] Mechelli, A; Price, CJ; Friston, KJ; Ashburner, J. Voxel-based morphometry of the human brain: methods and applications. *Curr. Med. Imaging Rev*. 2005 Jun; 1(2): p. 105-113.

[136] Deutsch, GK; Dougherty, RF; Bammer, R; Siok, WT; Gabrieli, JD; Wandell, B. Children's reading performance is correlated with white matter structure measured by diffusion tensor imaging. *Cortex*. 2005 Jun; 41(3): p. 354-363.

NEWS

In: Biochemistry and Molecular Biology Coming of Age
Editor: Mary Miu Yee Waye ISBN: 978-1-63321-815-4

Chapter 10

CONFERENCE REPORT OF 2010 INTER-UNIVERSITY POST-GRADUATE SYMPOSIUM

Edwin Chan
Department of Biochemistry, The Chinese University of Hong Kong,
Shatin, N.T. Hong Kong

The Department of Biochemistry, The Chinese University of Hong Kong hosted the first Inter-University Biochemistry Postgraduate Symposium on 15 May, 2010. Generously sponsored by The University Grants Committee of Hong Kong and Chung Chi College, The Chinese University of Hong Kong, this event was jointly organized with sister departments at HKU and HKUST with an aim to promote inter-institutional and inter-disciplinary communications among postgraduate students and faculty members. Over 180 postgraduate students and faculty members attended.

To mark the opening of the symposium, Prof. Zihe RAO, President of Nankai University and a leading figure in the field of protein structural biology, delivered a featured lecture on "Structure-based anti-viral research". More than 70 students presented their research findings at this inaugural symposium.

Further details can be found: http://www.bch.cuhk.edu.hk/bch2010/index.html

Scientific Program

8:15am – 9:00am	Registration + poster setup
9:00am – 9:15am	Opening remarks
9:15am – 10:15am	Plenary talk (Prof. Zhihe Rao, Nankai University, China)
10:15am – 11:00am	Tea reception + student poster session
11:00am – 12:20pm	Student stage presentation session I (4 presentations)
12:25pm – 2:30pm	Buffet lunch + student poster session
2:30pm – 3:50pm	Student stage presentation session II (4 presentations)
3:50pm – 4:30pm	Tea reception + student poster session
4:30pm – 5:50pm	Student stage presentation session III (4 presentations)
6:00pm – 6:15pm	Closing ceremony + award presentation

Platform Presentation Award:
Ng Ka Leung Andy (CUHK)
Pan Lifeng (HKUST)
Tsoi Ho (CUHK)
Poster Presentation Award:
Chen Jia (HKUST)
Chu Kit Hang (HKU)
Lei Ieng Lam (HKU)
Li Xiuling (HKUST)
Tam Chung Nga (HKU)
Tan Zhijia (HKU)

Selected abstracts are listed in Appendix 1.

In: Biochemistry and Molecular Biology Coming of Age
Editor: Mary Miu Yee Waye
ISBN: 978-1-63321-815-4

Chapter 11

Launching Theme-Based Research in Hong Kong

Anthony Chan
Research Grants Council of Hong Kong

An HK$18 billion Research Endowment Fund was established with allocation from the Hong Kong Government. HK$4 billion of the Fund is set aside for Theme-based Research which will support projects on themes of a longer-term nature and strategically beneficial to the development of Hong Kong. A Steering Committee, with membership consisting of the academia, representatives from different sectors of the community and the Government, and the Chairman of Research Grants Council (RGC) of Hong Kong, was set up with the major charge to identify the themes.

To launch the Scheme, funding will be focused initially on a small number of themes most beneficial to Hong Kong's strategic development and where the local academia have a competitive advantage, says RGC Chairman Professor Roland Chin.

Professor Chin points out that the research themes should aim to find solutions to important problems that are relevant to Hong Kong. The Theme-based Research Scheme's Principal Investigators must be academics from University Grants Committee (UGC)-funded institutions. It is hoped that the initiative will galvanize our strength and interest to work on high impact cutting edge research projects, and at the same time create state-of-the-art teaching materials and educational programs, as well as provide quality training for our research students.

Professor Chin says that the chosen themes are "Promoting good health", "Developing a sustainable environment" and "Strengthening Hong Kong's strategic position as a regional and international business centre".

Projects under the theme "Promoting good health" should aim to make Hong Kong a healthier society whose capability to prevent epidemics will set international standards. Research can focus on the fundamental life sciences, disease prevention and control, Chinese medicine, healthy living, aging, and health care policy, etc. Moreover, the Government's strategy of promoting medical services as one of the six priority economic areas is in line with this research theme, which is also the research agenda of other countries like Australia, Singapore and the Chinese Mainland.

"Developing a sustainable environment" means creating a sustainable and environmental-friendly development model for Hong Kong. Research on this theme may cover areas such as pollution prevention and control, sustainable energy, and recycling technology. This theme is of great importance because not only does it facilitate cooperation with the nearby regions and is ranked high on the research priority of governments and the academic community internationally, but it is also related to environmental industry which is one of the six priority economic areas identified by the Hong Kong Government.

Finally, the research theme "Strengthening Hong Kong's strategic position as a regional and international business center" can stand Hong Kong in a good stead to benefit from the rise of the Asian economy and enhance its status as a business center both regionally and internationally. Financial services and trading, logistics have always been two of the four traditional industries driving the Hong Kong economy. In light of the intense competition from the neighboring Asian cities, this research theme can contribute to Hong Kong's strategic economic development by strengthening our knowledge base in areas like risk management, wealth management and corporate governance, competitive models between Hong Kong and other Asian economies, and financial technology.

Professor Chin reiterates the importance of pooling together the best minds and resources to come up with the most beneficial solutions, such as engaging people from different academic domains to work on issues with major impact on Hong Kong. For instance, research on improving Hong Kong's air quality can involve researchers from many academic areas including mathematics, chemistry, engineering, and medicine, as well as academics in social sciences, law and economics. This can elevate the research impact of the Hong Kong academia to the next higher level.

Although the Scheme supports research of the UGC-funded tertiary institutions, overseas and Mainland scholars are encouraged to participate in the research projects. This will not only facilitate knowledge exchange between local and overseas researchers, but will encourage student participation from around the world, thereby enriching the research experience of the students studying in Hong Kong.

The RGC has called for white papers from local university researchers in early 2010, which will be used as a basis to identify the key research topics under each of the chosen themes. Once the key research topics have been identified, the RGC will invite proposals from Hong Kong researchers in September / October 2010. Successful research proposals will be awarded in 2011 after international peer review.

In: Biochemistry and Molecular Biology Coming of Age
Editor: Mary Miu Yee Waye
ISBN: 978-1-63321-815-4

Chapter 12

CONFERENCE REPORT OF 2011 HONG KONG INTER-UNIVERSITY BIOCHEMISTRY POSTGRADUATE SYMPOSIUM

Bryan Ho, Julian Tanner and Dong-Yan Jin
Department of Biochemistry, The University of Hong Kong

With the aims to facilitate a better understanding of current biochemical research as well as to promote collaborations and interactions among postgraduate students and young scientists in the three major universities in Hong Kong, the Department of Biochemistry, The University of Hong Kong (HKU) hosted the second "Hong Kong Inter-University Biochemistry Postgraduate Symposium" on 11th June, 2011. This event was jointly organized by the Biochemistry Program/Section of the Chinese University of Hong Kong (CUHK) and the Hong Kong University of Science and Technology (HKUST). Two highly esteemed international scientists were invited to give keynote lectures in this symposium – Professor Takashi Fujita from the Kyoto University and Professor Ben Berkhout from the University of Amsterdam.

In his lecture on "Antiviral innate immunity induced by RIG-I-like receptors", Prof. Fujita reviewed the biochemical and biological roles of a subfamily of DExD/H box RNA helicases consisting of three members named RIG-I, MDA5 and LGP2. These RIG-I-like receptors function as sensors of viral nucleic acids to boost host innate immune responses including the production of type I interferon. Recent findings in relation to the activation of mitochondrial adaptors by RIG-I and the stimulation of RIG-I by RNA molecules of some plant species were discussed. Prof. Fujita also shared his views on the pursuit of ground-breaking research based on crazy ideas, non-canonical approaches and hard work.

In his lecture on "Insight into HIV-1 drug resistance mechanisms through virus evolution", Prof. Berkhout presented examples in which HIV-1 escape from the action of peptide and RNA drugs, including peptide inhibitors of fusion and HIV-1-targeting short hairpin RNAs. Peptide inhibitors of fusion such as Fuzeon can block the entry of HIV-1 into host cell. Through virus evolution, the mechanisms of drug resistance were delineated, leading to the development of next-generation therapeutic agents. It was shown that HIV-1 cannot escape from either peptide mimetics with an introduced salt bridge that masks the charged residues or the combined use of four potent short hairpin RNAs. Prof. Berkhout also gave comments on future directions of antiviral therapy and gene therapy for HIV-1.

The symposium was well attended with more than 150 participants with a fairly even mix of students from HKU, CUHK and HKUST. Students gave oral and poster presentations across a diverse range of biomedical research relating to biochemistry. Three best oral presentation awards and six best poster presentation awards were given to outstanding presenters (Table 1).

Concurrently, three awards from the Hong Kong Society of Biochemistry and Molecular Biology also go to the best oral presenters.

This symposium successfully provided a platform for interaction among students from different institutions in Hong Kong, and enhanced inter-university understanding and collaboration.

The organizers thank the Hong Kong Research Grants Council and other sponsors for their support of this event. Further details of this conference can be found at: http://www0.hku.hk/biochem/postgraduate/symposium2011/index.html

Table 1. List of best presentation awardees

Awardee	University	Title of presentation
Wei Liu	HKUST	Dynamic scaffolding of the INAD signaling complex via conformation-coupled protein disulfide redox potential cycling (oral)
Jacques Pak Kan Ip	HKUST	α2-chimaerin controls neuronal migration and lamination of cerebral cortex through regulation of CRMP-2 (oral)
Ho Tsoi	CUHK	Expanded CAG RNA trigger cell death by disrupting nucleolar function in polyglutamine diseases (oral)
Kit Hang Chu	HKU	Sox10 is essential for inner ear gangliogenesis (poster)
Ya Lun Zhang	HKU	The role of Pax6 in β-amyloid toxicity and Alzheimer's disease (poster)
Chi Shing Yu	CUHK	Revisiting co-evolution theory of the genetic code from a whole-genome perspective (poster)
Ren Tian Wu	HKUST	Cdt1p, through its interaction with Mcm6p, is required for the formation, nuclear accumulation and chromatin loading of the MCM complex (poster)
Wing Yip Tam	CUHK	Redundant functions of LIM-homeodomain transcription factors Lhx1 and Lhx5 on postnatal development of cerebellar Purkinje neurons in the mouse (poster)
Sam Kit San Yuen	HKU	Adenoviral delivery of RNA decoys induces silencing of an Epstein- Barr virus-encoded antiapoptotic microRNA and restoration of PUMA expression in nasopharyngeal carcinoma cells (poster)

Hong Kong Inter - University
Biochemistry Postgraduate Symposium 2011
Cheung Kung Hai Conference Centre, HKU
11 June, 2011
The Research Grants Council

PERSPECTIVES

In: Biochemistry and Molecular Biology Coming of Age
Editor: Mary Miu Yee Waye
ISBN: 978-1-63321-815-4

Chapter 13

CHAOS THEORY IN ASTROPHYSICS, NATURE, GENETICS, AND HUMAN CANCER

Jessica Li**[*]**
School of Biomedical Sciences,
The Chinese University of Hong Kong, Shatin, N.T. Hong Kong

ABSTRACT

Chaos theory predicts the causation and possible outcomes of systems in nature and explains the ever increasing entropy of nature. By observing nonlinear data, it is utilized to find underlying order in apparently random data. Human cancer and diseases caused by aneuploidy, gene, and chromosome anomalies may be analyzed using this model as well as quantum mechanics and the universe as a whole. Chaos theory is applied from the very small world of the subatomic to the very large world of the cosmos. It can be used to observe many very different aspects of nature.

I. INTRODUCTION

The chaos theory states that the behavior of the universe is unpredictable with chaos being a nonlinear function. Chaos explains the irregularities in the universe and in the world including natural disasters, the weather, evolution and orbits of planets and stars. Two important terms to note regarding the chaos theory are *the butterfly effect* and *sensitivity to initial conditions*. The butterfly effect emphasizes that the incredibly tiny magnitude of deviation in initial conditions to create enormous effects in a system.

The butterfly effect explains that a flap of a butterfly's wings can cause a hurricane elsewhere, possibly on the other side of the world. The flap is analogous to the tiny magnitude of deviation in initial conditions and the tornado represents the resultantly large magnitude of the wing's flap.

---

[*] Corresponding Author: Jessica Li, Email: jess.jsli@gmail.com Keywords: chaos theory, astrophysics, cancer, genetics.

The chaos theory emphasizes the importance of sensitivity to initial conditions. A tiny deviation in initial conditions causes an enormous effect in the outcome.

Ia. Demonstration of Chaos Theory

Chaos can be tested by using a double pendulum which is made up of two rods as shown in Figure 1. Two rods are joined together at a pivotal point. The pendulum's trajectory movements are predictable from the start but progressively become from less predictable to chaotic. For example, if two double pendulums are set side by side, their traced movements would be drastically different after a certain amount of swings, demonstrating *sensitivity to initial conditions.* A double pendulum's movements are affected by a few factors: air resistance, attached weights, length of the pendulum rods, the height at which the pendulums are released, and the force at which the rods are swung. All these factors contribute to the pendulum's change in direction and velocity just as various factors affect any system in nature.

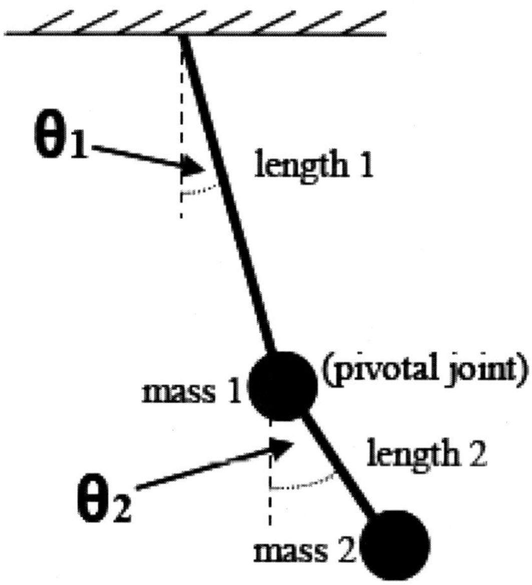

Figure 1. Double *Pendulum* (modified from [8]).

Ib. Nature, Pollock, and the Chaos Theory

Jackson Pollock was an artist from the early to mid 1900s. Pollock's works of drip paintings created one of the earliest studies of the chaos theory. Using fractal geometry, Pollock's drip paintings were analyzed using computer image processing techniques and then applied to the chaos found in nature. The fractals in his paintings resemble the branches of a tree. These fractals as a whole may look chaotic, but when a single portion of a branch or fractal is magnified, order and organization can be clearly identified [12]. This can be shown through analyzing Mandelbrot sets as shown below in figure 2. Mandelbrot sets are derived from a single simple mathematical formula, but its pattern branches off into chaotic fractals.

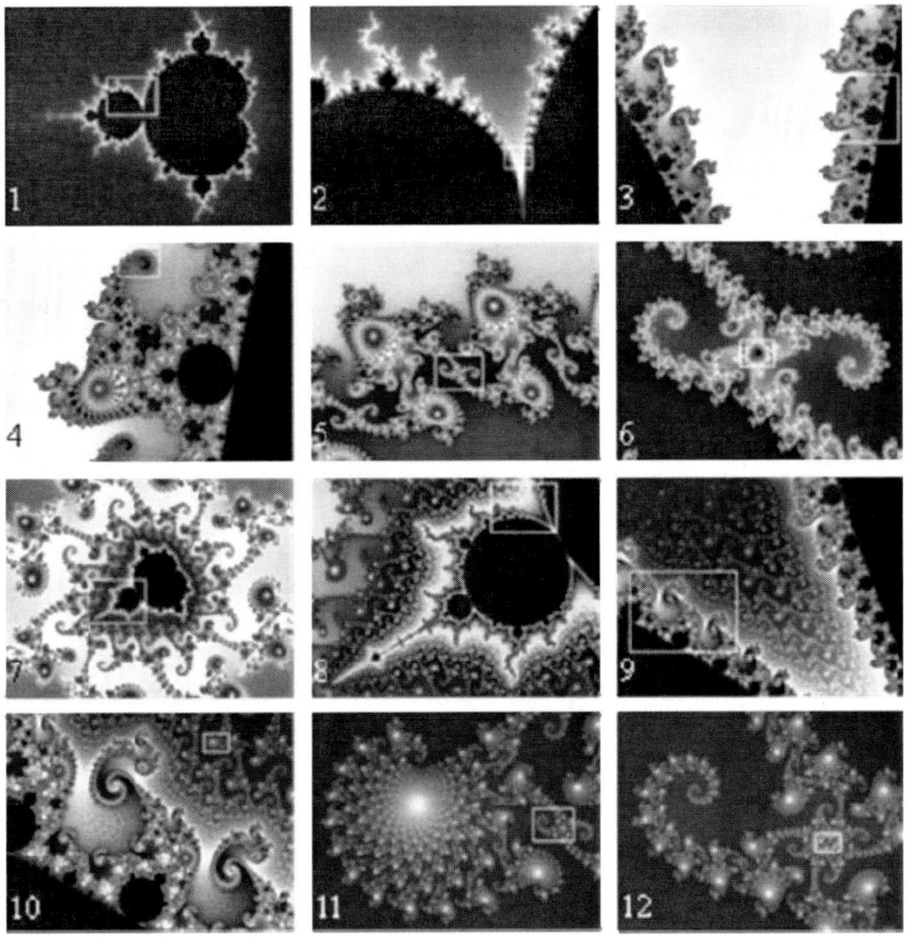

Figure 2. *Mandelbrot set* (1). Magnification of fractals are demonstrated using a Mandelbrot set. The numbers marked on the bottom right hand corner of each picture in the set indicate the step of magnification. The boxed area in each step shows the portion of the image that is magnified in the next image. Step 1 is the most zoomed out while step 12 is the furthest magnified of the original image of step 1. Throughout the magnifications of this single image (step 1), images of higher magnifications can be related back to previous steps of less magnified images. For example, step 8 looks very much like the left portion of step 1 and in steps 4, 5, 5, 9, 10, 11, 12, the theme of curled fractals is apparent throughout magnifications. "A small part of the structure looks very much like the whole" [12].

II. Bringing It All Together

IIa. Uniting Astrophysics and Biology

In theoretical physics and astrophysics, the universe began as a singularity which is an infinitely dense point of matter compressed into a small area. This singularity then expanded through the Big Bang when the universe was born. The Big Bang, like an egg and a sperm in biology gave birth to the beginning – the start of all possibilities of life.

Before the universe expanded, the universe was supersymmetric at the Planck energy, meaning that all four fundamental interactions or forces of the universe: electromagnetic force, strong interaction, weak interaction, and gravitational force were of equal magnitude as shown in Figure 3. This equality is analogous to the development of an embryo with unprogrammed or undifferentiated cells. All cells are the same in the undifferentiated stage with "absolute chromosomal conformity" [3]. In this supersymmetric stage of the universe and undifferentiated state of the embryo, no useful functions can be performed. "Life as we know it could not exist in this perfect state" [6].

After the Big Bang symmetry breaks and the four fundamental forces gradually diverge and become unbalanced, deviating to different magnitudes which allows for the formation of celestial objects (refer to Figure 3). Different types of particles with different functions and masses as shown in the Standard Model are created. The universe continues to expand and grow, spawning new bodies and celestial objects such as galaxies, stars, planets, and blackholes. Different types of galaxies with spiral, elliptical, and irregular shapes form along with a variety of stars – neutron stars, protostars, binary stars, variable stars, etc. At this point, specialization of cells in the embryo occurs. Cells differentiate into heart, brain, nerve, skin, bone, muscle, etc. to allow a self sustaining human body to form and function on its own.

Differentiation of human cells and the imbalance of the four fundamental forces state an attempt to maintain homeostasis and functionality in their systems of nature. For example, too many or too few blood cells in the human body would hinder the body's function causing diseases such as leukemia, leucopenia, and anemia just as too large a magnitude of the strong nuclear force would cause the destruction of a star into a black hole before it even formed.

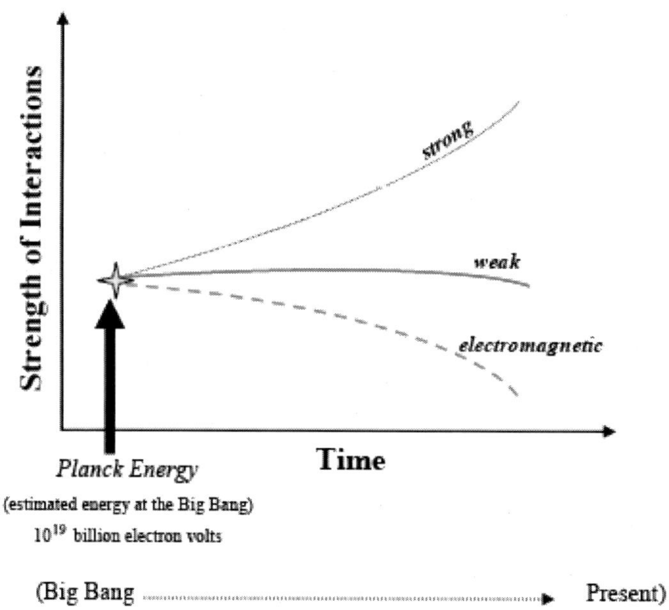

Figure 3. Supersymmetric theory (modified from [6]) "The strengths of the weak, strong, and electromagnetic forces are quite different in our everyday world. However, at energies found near the big bang, the strengths of these forces should converge perfectly" [6].

Expansion and inflation of the universe speeds up exponentially just as chaos and entropy accumulate exponentially. Using WMAP, scientists determined the universe as a mostly uniform body; however, when analyzing the formation and destruction of celestial bodies, there is much disorder involved. The quantum theory states that electrons and subatomic particles pop in and out of existence according to Heisenberg's Uncertainty Principle meaning that in a way, there is no absolute way to predict the absolute location of a specific molecule. Moreover, random events of galaxies' gravitational forces ripping each other apart, black holes sucking in debris, and supernovae are a combination of death and birth all in one happening billions of times simultaneously across the universe. So, according to the butterfly effect - does one event trigger another halfway across the universe or are the cosmos simply just considered chaotic without any means of order?

In the universe, the amplitude of irregularities in the cosmic microwave background is known as the Q factor. According to Kaku, if the Q factor in the universe were larger, "matter would have condensed earlier in the history of the universe into huge supergalactic structures" meaning that there would be enormous black holes.

This irregularity can be compared to the malignancy of cancer; both abnormally large black holes and cancerous cells would consume and displace nearby galaxies and healthy cells, respectively and cause chaos [6].

Following the path of nature, entropy increases exponentially and when chaos accumulates so much so that the human body cannot handle it, the body's cells may rebel and stray from its original functions. This chaos can be defined as human disease or cancer.

IIb. Evolution and Human Cancer Modeled After Chaos Theory

Evolution's natural selection is the process by which certain traits get selected and passed down through generations of species to create organisms that are more ideal or likely to adapt and survive. In chaos theory, natural selection can be viewed as finding organization in an otherwise chaotic gene pool. Genetic variation within a species can also be identified with chaos theory. In genetics, each specie has a specific order and number of genes that one specie distinguishable from another. Each specie has its own genome. Furthermore, each organism within a specie functions and looks similar but within the individual organism is genetic variation which makes each organism unique from another, even within a specie. For example, within the human race each person is unique due to natural genetic variation from the swapping and combining of the genes of each person's biological parents.

In the development of biological systems or organisms, all initial states of possible systems have equipotentiality, meaning that they all have equal opportunities to emerge or evolve into more complex systems. These complex biological systems would then have the ability to self-organize and regulate. According to Janecka, systems can swing in and out of chaos. In the healthy zone of chaos, systems are still stable and able to regulate themselves to maintain homeostasis. When a system swings out of that healthy zone into the outer edge of chaos, the amount of entropy is too much for the system to handle and the system's survival is being greatly threatened and may fail. This is what happens when a healthy human being's system gets pushed into overwhelming chaos. A person's body would be no longer able to defend and regulate itself in the outer edge of the zone of chaos. This threat to a human body's system is known as the devolutionary phase, including disease, cancer, or aging.

Aging represents an increase in entropy which means the system is more susceptible to diseases and mutagenic agents. Functionality depends on the amount of entropy in the system. The more entropy there is the lower the functionality and vice versa [5].

Human cancer and disease arise as a result of the accumulation of carcinogenic materials and self propagating destructive cycles through many cell and human generations. "...there is growing evidence that genomic instability and the resulting aneuploid genotypes and global gene expression changes have a significant role in the onset of cancer." Genomic instability and chaos is a result of these accumulations. Carcinogens damage genetic pathways of normal cells by mutating their genes. These mutations then activate oncogenes and which deactivate suppressor genes. This causes the natural selection to produce resistant and cancerous cells through disruption of cell cycle controls. Comparing normal cells to cancer cells, normal ones are more orderly than cancerous ones thus supporting the model of chaos theory that higher entropy represents lower functionality. Normal cells have mitogens or natural growth factor stimulators and proteins to control cell growth; both of which enable cells to naturally enable homeostasis. Healthy human cells are always diploid; also cells of the same type and function are of relatively the same size and shape operate accordingly with their programmed function. Cancerous cells on the other hand are aneuploid, meaning they have either gained or lost whole or segments of chromosomes. They may also have developed auto-activation of signal transduction pathways, deactivated of tumor suppressor genes, and resistance to apoptosis which are malignant to a body's system. Loss of normal functions allows such cells to operate in a sense of autonomy by developing new traits and behaviors. This causes failure of the system to operate and maintain homeostasis as a whole. These cells are also chaotic due as they develop their own shapes, sizes, growth and metabolism rates because of aneuploidy. Due to aneuploidy, the amount of genetic content in each cell widely varies; one aneuploidy cell may contain double the normal amount of DNA while another aneuploidy cell may contain only half the normal amount of DNA.

These discrepancies then hinder the cells' ability to repair or operate normally [3]. However, these chaotic mutations might also be a mechanism for generating more desirable genomes that may be due to ever changing environmental conditions. For example, some transposable elements are known to help the fruit flies to spread out of Africa while others were found to play a role in adaptation to temperate climates [4]. Recent advances in second generation high-throughput sequencing and genome-wide comparative analyses of different human populations, as well as between species (such as human versus chimpanzee) have helped us understand the relationship between diseases and selection.

The 1,000 Genome Project, launched in January 2008, aims at creating a complete and detailed catalogue of human genetic variations of genomes of about 2000 unidentified people from about 20 populations around the world using next-generation sequencing technologies. This catalogue will be a useful tool for studying regions under selection and understanding the underlying processes of mutation and recombination. The results of the 1,000 Genome Project will be freely and publicly accessible to researchers worldwide so that scientists from all disciplines will be able to contribute to the analyses and follow up with additional studies of variation in particular regions.

Other more detailed descriptions of individual genomes of famous scientists such as James Watson and Craig Ventor also revealed how one individual genome can harbor so many variations - concordant with what one would expect from the chaos theory [11].

CONCLUSION

The chaos theory created a major impact on science and is able to unite very different categories of science such astrophysics and biology to observe human cancer and nature in a new and unique perspective in modern science. Just as physicists are trying to find a grand unified theory for uniting quantum mechanics and relativity, biologists are trying to find a sense of organization and unity in genetics to understand and treat cancer. The search in molecular biology has barely begun.

REFERENCES

[1] Beyer, W. Mandelbrot set [online]. 2007 [cited 2010 Aug 26]. Available from: http: //commons.wikimedia.org/wiki/User:Wolfgangbeyer.

[2] Check, E. James Watson's genome sequenced [online]. 2007 [cited 2010 Aug 30]. Available from: http://www.nature.com/news/2007/070528/full/news070528-10. html

[3] Duesberg, P. Chromosomal Chaos and Cancer. *Scientific American.* 2007; 53-59.

[4] González, J; Lenkov, K; Lipatov, M; Macpherson, JM; Petrov, DA. High Rate of Recent Transposable Element–Induced Adaptation in Drosophila melanogaster. *PloS Biol.* 2008; 6.

[5] Janecka, IP. Cancer control through principles of systems science, complexity, and chaos theory: A model. *International Journal of Medicinal Sciences.* 2007; 164-73.

[6] Kaku, M. *Parallel Worlds.* USA: Anchor Books; 2005.

[7] Levy S; Sutton G; Ng PC; Feuk L; Halpern AL; et al. The Diploid Genome Sequence of an Individual Human. *PLoS Bio*l. 2007; 5.

[8] Oxford University Press [Internet]. USA [updated 2008 Sept 25; cited 2010 Aug 25]. Available from: http://pages.physics.cornell.edu/~myers/teaching/Computational Methods/ComputerExercises/DoublePendulum/doublependulum.html

[9] Schneider, BL; Kulesz-Martin, M. Destructive cycles: the role of genomic instability and adaptation in carcinogenesis. *Carcinogenesis.* 2004; 25: 2033-44.

[10] Shreeve, J. *The Genome War: How Craig Venter Tried to Capture the Code of Life an*d *Save the World.* Knopf; 2004.

[11] Spencer, G. International Consortium Announces the 1000 Genomes Project, EMBARGOED [online]. 2008 [cited 2010 Aug 30]. Available from: http://www. 1000genomes.org/files/1000Genomes-NewsRelease.pdf

[12] Taylor, RP. Chaos in Pollock's Chaos. *Scientific American.* 2002; 117-121.

PATHS TO SUCCESS

In: Biochemistry and Molecular Biology Coming of Age
Editor: Mary Miu Yee Waye
ISBN: 978-1-63321-815-4

Chapter 14

FORGET ABOUT COMPETITION AND STRIKE FOR THE BEST

Masayo Kotaka
Department of Physiology, Li Ka Shing Faculty of Medicine,
Hong Kong University

Back then, The Chinese University of Hong Kong still provided four-year bachelor degree. Dr. Kotaka was admitted to the department of biochemistry through the JUPAS system and graduated with First Honors after 4 years. She has a long history of studies with Professor Waye. At the summer break of her first year, Dr. Kotaka hoped to engage in some experiments in the laboratory, and after a meeting with Professor Waye, her wish was granted. Dr. Waye was also her advisor when she studied for her PhD degree. Dr. Kotaka is now working on the studies of the structures of protein using X-ray. By observing how the enzyme works in the body could help develop new antibiotics and anti-virus medicine.

Many mainland students rush into Hong Kong to study for their Master or Doctorate degree in recent years. In Dr. Kotaka's team, about 70-80% of the research team members are from Mainland China. It has come to Dr. Kotaka's attention that most locals students tend to choose their way based on the current economic situation: to continue studies or to work. A favorable economic condition usually encourage the local students go to work after graduation. But during an economic downfall with fewer job opportunities, students tend to pursue for further studies. Hong Kong's recent economic performance is relatively stable, so it seems that less students apply for the Master and Doctor degrees.

There are quite a number of talented students in Mainland and surely they will cause competition with the locals. But the local students have their own advantages. They often have a higher flexibility and adaptability. With a strong learning motivation, most of the local students are persistent and patient for the different stages of the experiment. The journey to success is always long and slow, numerous amendments are needed. When come to this plight, most of the local students are willing to re-adjust and think about their strategies. From Dr. Kotaka's own experience, some of the students, including the ones from the mainland, do not take the failure in experiments very well, the adaptability and adjusting ability are relatively low. The eager for achieving success in a fast and easy way would often hinder the experiments. The disappointment brought by failure is no stranger to the scientists, only by adjusting oneself could embrace more challenges.

The support from government for the local students is essential. China has such a high population that even a small portion of in-coming students could easily outnumber the local's. A higher priority given to the local students to pursue postgraduate studies could give them more chances. In USA, universities of each state would retain higher priority or deduction in tuition fee for the local students as an encouragement. Since the exam systems of China and Hong Kong differ, it is reasonable to secure some positions for the local students.

Local university could also step in to encourage students to pursue further studies. When studying for the bachelor degrees, many students do not have a clear plan for their future and their GPA may hinder their further studies. If they could be exposed to more types of experiments and get more chances of practical experience, it would definitely stimulate their interests.

Students of China and Hong Kong are not competitors. The best way to study is to be responsible for one's own research topic. The failure would turn into a meaningful experience which could pave a smooth road to later successes.

In: Biochemistry and Molecular Biology Coming of Age
Editor: Mary Miu Yee Waye
ISBN: 978-1-63321-815-4

Chapter 15

THE PRICELESS GIFT FROM RESEARCH AND EXPERIMENTS

Kit Ray Ng
Department of Pathology, Li Ka Shing Faculty of Medicine, Hong Kong

When he was still studying for the bachelor degree for biochemistry in CUHK, Professor Ng has heard about Professor Waye's research even he did not study any of her courses. Professor Waye then became his counselor when he pursued his master degree. She is an excellent counselor who is very concerned about her student's future and research direction. Professor Waye had provided much support and wrote a very strong recommendation letter for Professor Ng's application for doctorate degree in Cambridge University. For now, Professor Ng's research covers stem cells and blood cancer, including make blood cells with embryonic stem cells.

The recent trend shows that many mainland students come to Hong Kong for further studies. Facing the keen competition, many local students stop continuing their studies after graduation. But that is not the sole reason. Researches and experiments often require long period of time and lots of patience, and they are the influencing factors. However, the learning process in master or doctorate degree could enable students to get more knowledge of techniques and discover more about themselves. These would be very valuable for the future jobs. The experiments of undergraduates could easily be taken from a limited reference from text books, but in reality, there are no limitation in for researches and experiments. Students need to self-motivate for learning more knowledge and get the skills of processing tons of data. Experiments could cultivate patience and carefulness, also the ability to face failure. The world of science has no boundaries and scientists around the world are competing to make the first discoveries. All experiments, like what Professor Ng told his students, fail most of the time. The failed experiments teach students a lesson: adjust themselves when facing failure and pressure.

During his exchange period in Cambridge University, Professor Ng found that quite a lot of scientists exchange opinions or share latest information during lunch time or tea time. These relaxing gatherings actually could help a lot in the research progress. Hong Kong students always need to face pressure from parents and societies when they choose to develop a career in the research sector. However, western societies hold a high respect for the scientists and the scholars. Money is not the measurement of social status, gaining knowledge is the truly great achievement.

The joy of successful experiments should not be shared only within the laboratories. By promoting the fruit of researches, we could educate the public more about the value of pursuing scientific research, it could also encourage more youth to take science as their careers. Universities now take an active role to promote the success of researches and experiments to secondary school students in a more simple way; open lectures and newspaper articles also get more public attention, the distance of the labs and the society is shortened. In recent years, the government invested more resources on the researches of university and medical laboratories to improve the public health conditions. The resources in university are, sadly, limited. If the government could provide relatively more job opportunities, more top scientist would be recruited.

Hong Kong has a close connection with the international society. It has also become a trend to cooperate with the mainland firms, many bio-technology firms set up branches in Hong Kong and China. With the mature international network, Hong Kong could be the bridge to promote China's research results. Scientific researches may not create wealth, but the priceless knowledge could definitely push the world to be a better place for humankind.

In: Biochemistry and Molecular Biology Coming of Age
Editor: Mary Miu Yee Waye
ISBN: 978-1-63321-815-4

Chapter 16

CRAVE FOR IMPROVEMENTS AND STEP FORWARD

Ming Yuen Lee
Institute of Chinese Medicine, University of Macau

After getting his Bachelor Degree of Biology at City University of Hong Kong, Professor Lee Ming Yuen worked for half a year. He was then given an opportunity to join Professor Waye's research team at the Chinese University of Hong Kong and became a research assistant. When compared with the jobs at commercial labs, the half-year experience as a research assistant was more appealing and delightful. So, Professor Lee chose to stay at the Chinese University of Hong Kong to study for his doctoral degree. He is now teaching at the Institute of Chinese Medicine of University of Macau, specializing at the research about Chinese medicine and molecular pharmacology. Current medicines may not tackle illnesses like cerebral hemorrhage or neurodegenerative disease well, as a result, Professor Lee tries to find a new cure from the investigations of Chinese medicine and high throughput screening.

Scientific researches consist of challenges different from those in the testing services of commercial labs. The experiments in commercial labs usually are in a quite mature stage, reliable and predictable results could always be generated. On the contrary, the managements and processes of experiments are more important. Unlike the experiments of commercial labs, scientific researches could start at a brand new field of significance and fun. The research results could also be published at international magazines and intrigue the others. It cannot be doubted that the research process is tough, failure could keep coming, and all problems could not be solved in a short term. But given enough time and resources, accompanied with the craving to know the reason of failure, `Giving up ' is never a good choice.

The city we are living is changing, just like Asia and the world. Even in a close distance, Hong Kong and Macau still have the cultural difference and varied concepts on the resources allocations. Since the end of the monopoly system, the Macau government has put more resources on teaching and researching, resources and funds are quite evenly distributed among the scholars. In Hong Kong, the basic infrastructure are developed quite well, universities could share facilities and thus making it convenient for scientific research. These two places have their own advantages. The prosperous economic condition in Macau let her citizens enjoy a higher standard of living, and most graduates will take up a career in tourism or gambling businesses. Jobs related to scientific researches are less favored. Most of the students pursuing mater degrees or doctoral degrees in Hong Kong or Macau are from Mainland China.

The demand for researchers actually has never decreased. The outbreak of infectious diseases, food safety and pollution of natural environment are becoming the main concerns of the public. They are related to the basis of natural science, and researchers are the one who could lead the way. The mobility of researchers are quite high as results and technique exchanges are in placed at a high priority. Professors and students could go around the world for exchanging experiences. The results of a research start from an idea, followed by a series of experiments. `From none to all' could generate the biggest pride.

Researchers need to always push themselves further for improvements. They have to face the competitions from global scholars and researchers, the passion for science support them through all the pressures. The road of science is pointing upwards, only the never-stopping climb and struggle could let you see more and clearer. To achieve the goals, one must never slack off.

In: Biochemistry and Molecular Biology Coming of Age
Editor: Mary Miu Yee Waye
ISBN: 978-1-63321-815-4

Chapter 17

Try Hard, and Curiosity Will Lead to Success

Hoi Yeung Li
School of Biological Sciences of Nanyang Technological University

Graduated from Polytechnic University, Professor Li Hoi Yeung gained his M. Phil degree at The Chinese University of Hong Kong. Dr Mary Waye, who has questioned him at his interview for the application later became his Ph.D. advisor. Currently Professor Li teaches at the School of Biological Sciences of Nanyang Technological University.

The current research field of Professor Li is about cell division, the origination of multi-cell organism like human. It is a field tracking from the origination for an answer every man tries to seek. Scientific researches aims at discovering the unknown and solve the mystery, this rule applies in Hong Kong, also at Singapore where Professor Li is working. The answers we are looking for lay in the hands of the scientists and researchers. Their experience, their talent, their determination to search for the root, will eventually lead us to the answer.

Professor Li believes that there is a 99.9 % failure rate for experiments, with only a slim chance to succeed. Just like any other normal people, he is troubled by pressure and the decision whether to give up or not. Professor Li enjoys cooking very much, and he finds some similarities between cooking and doing experiments. Failure only means a new chance for trial and improvement, while determination can bring the fruit of success. Curiosity and the fun during experiments also drives him more. Professor Li even joked that failure is very normal for him, he could embrace it peacefully. If he could get a satisfying result from a challenging experiment, the pleasure could not be traded by money.

The seeds we plant today are the fruits we pick in the future. Researches are the key to changing the future. The technology we used today might originate from a research result of more than 30 years ago. When the society and the world stop all the research efforts, we are stuck at the current age. Biotechnology exists to explain the unknown field. We now rely on the technology to bring us breakthrough, so we could extend our now limited knowledge about cancers, or cancers will still take lives away. When human stop to innovate, we are not moving on. And when students choose to ignore their talents and take another way, they may have lost the chance to make changes for the future generations.

Interest drives people to work and succeed. Most of the research projects in the society are not continual. Once the mission is accomplished, another assignments will come. It is totally different on the road of researches. One research topic will be the centre of all the researches in researchers' career, the pressure and pride will accumulate through working hard on it too. Not all men has the ability and the endurance for research, if students possess the talents, they should really consider contributing for scientific research. Material life will not be abandoned; a different field of work could still bring you success.

To students or people who have started working, doing research requires a clear target. Persistence, patience, curiosity and sense of honor could make people strike for the best, contribute for mankind and our society.

In: Biochemistry and Molecular Biology Coming of Age ISBN: 978-1-63321-815-4
Editor: Mary Miu Yee Waye

Chapter 18

BACK TO THE ROAD TO RESEARCH

After graduating with a bachelor degree of biochemistry after 4 years of studies, professor Lam Wai Yip chose to still stay in the Chinese University of Hong Kong (CUHK) to pursue his M.Phil degree of biochemistry, it was also the very time that Dr Waye came to CUHK. Professor Lam then asked for her to be his supervisor and joined in a project that Dr Waye cooperated with the University of Toronto.

Professor Lam minored in computer science when he was an undergraduate; this knowledge enabled him to take up the bioinformatics-related work in Dr Waye's research project. More than that, his minored subject affected his choice of job, made him choose to work in the IT department of a bank after the graduation of the M.Phil degree. Commercial society is extremely efficiency-oriented and gives benefits a high priority. It was depressing that private time needed to be sacrificed in order to complete the workloads that come with so many quotas and deadlines. With a longstanding preference for research, Professor Lam started to think about resignation and going back to university for further studies He went through quite a lot of struggles, especially about the financial management. The difference between his current salary and the monthly remuneration did make him hesitate. But one should not be caged by the social rules, life is long enough to fulfill dreams and interests. Professor Lam then handed in his resignation letter and pursued the doctorate degree, which leaded him to cooperate with Dr Waye again in a research about using Chinese medicine to cure hepatitis.

Professor Lam's road is not flat without any bumping. He did not pass the final defense presentation as something occurred in his family. It was rather dispiriting that the thesis he worked hard for years could not pass. Lukily, he is supported by Dr Waye and his friends. So he soon returned to the `battlefield' and graduated after half a year. The support from family and friends is essential, but it is insistence makes a difference. The opposing advices of friends may have reasons; however, it is not wrong to insist your principles. There may not be instant returns and outcomes for a research, but patience and insistence will pay back.

Unlike other professors, Dr Waye seldom set up a task for students to finish, she put more emphasis on freedom and experiment, let her students set up their own topic to research for. Guiding them without intervention, she effectively stimulates the mind and the independent thinking ability of her students. Perhaps got affected by her approach, Professor Lam would not give his students too much homework too, he tends to focus on their own initiative to learn.

His previous research is mostly related to influenza virus like H5N1 or respiratory virus. In this year, he mainly focuses the genetic sequencing experiments about bacteria and viruses, and an experiment about micro-RNA. He wishes that Dr Waye would enjoy a fabulous live after environment.

APPENDIX: SELECTED ABSTRACTS FROM THE 2010 INTER-UNIVERSITY POST-GRADUATE SYMPOSIUM HELD IN HONG KONG

STRUCTURAL AND FUNCTIONAL STUDIES OF ZONULA OCCLUDENS-1

Jia Chen*[*1, 3], Lifeng Pan[1], Jiang Yu[1], Zhiyi Wei[1], Yanxiang Zhao[2] and Mingjie Zhang[1]

[1]Department of Biochemistry, Molecular Neuroscience Center, Hong Kong University of Science and Technology, Clear Water Bay, Kowloon, Hong Kong

[2]Department of Applied Biology and Chemical Technology, Hong Kong Polytechnic University, Hong Ham, Kowloon, Hong Kong

[3]Department of Nanoscience and Nanotechnology, Hong Kong University of Science and Technology, Clear Water Bay, Kowloon, Hong Kong, China

Intercellular junctions, including tight junctions (TJs), adheren junctions (AJs), and gap junctions (GJs) as defence barrier, association points and communication channels plays important role in polarized epithelial cell or cell polarization events such as directional migration. Zonula occludens-1 (ZO-1), one of the junction marker proteins, is also the first identified member of the PDZ family of proteins. ZO family, ZO-1/2/3, are proved to have critical functions in the formation and maintenance of intercellular junctions. But how they organize and possibly regulate junctional complexes need to be further investigated. In my PG seminar talk, I will present our research progress in studying two regions of ZO-1: the second PDZ domain (PDZ2), and the multi-domain segment containing the third PDZ, the Src Homology domain (SH3), and the guanylate kinase (GuK)-like domain.

[*] Email: chenj@ust.hk.

By using NMR, X-ray crystallography as structural approaches and biochemical methods such as isothermal titration, fluorescence polarity assay and etc, we found the PDZ2 domain of ZO-1 can form swapped dimer which generate a specific and regulatory Connexin43 (Cx43) binding groove. Another interesting finding about ZO-1 is that its third PDZ domain together with its following SH3 and GuK domains form a structural supramodule, which may provide functional information for its binding with TJ protein occludin, Y-box transcription factor ZONAB, and other junctional proteins. In summary, our studies have provided structural and functional indication on how ZO-1 modulates the junctional complexes and possible regulatory mechanism.

GLOBAL GENE EXPRESSION CHANGES DURING CHONDROCYTE ADAPTATION TO ER STRESS

Zhijia Tan*[1], Ben Niu[1], Kwok Yeung Tsang[1], Ian Melhado[1], Michael Zhang[2], Danny Chan[1] and Kathryn S. E. Cheah[1]

[1]Department of Biochemistry, Faculty of Medicine, the University of Hong Kong
[2]Cold Spring Harbor Laboratory, 1 Bungtown Road, Cold Spring Harbor, NY11742, USA

In humans, mutations in *COL10A1* trigger ER stress and result in metaphyseal chondrodysplasia type Schmid (MCDS). In the 13del transgenic mouse model with MCDS mutations, the Col10a1 transgene with a 13bp deletion is expressed in the hypertrophic chondrocytes (HCs). The mutant 13del proteins cannot fold properly in the endoplasmic reticulum (ER), thus inducing ER stress and trigger the unfolded protein response (UPR). The terminal differentiation of 13del HCs is impaired. These cells reprogram to alleviate ER stress and survive, including cell cycle re-entry and re-expression of markers characteristic of pre-hypertrophic chondrocytes (Tsang et al., 2007). To dissect the underlying mechanism of chondrocyte reprogramming and adaptation to stress, different regions in the wild type and 13del growth plates were fractionated for global gene expression profiling. Comparative transcriptomics between wild type and 13del HCs reveal novel ER stress inducible genes. Up regulation of genes expression related to oxidative stress and hypoxic stress pathways suggests that adaptation of HCs to ER stress in vivo involves cross- regulation by other stress pathways. Factors and regulatory circuitries that might be responsible for the chondrocyte reprogramming and ER stress adaptation will be predicted by bioinformatics approaches.

* E-mail: zjtan@hkusua.hku.hk.

Reference

Tsang, K. Y., Chan, D., Cheslett, D., Chan W. C. W., So C. L., Melhado, I. G., Chan, T. W. Y., Kwan K. M., Hunziker E. B., Yamada, Y. Bateman, J.F., Cheung K. M. C., Cheah, K.S.E. (2007). *PLoS Biology, 5*(3), 0569-0585.

Genetic Analysis of Zebrafish Hematopoiesis-Defective Mutant Thy-193

Xiuling Li*[*,1], *Jin Xu*[1], *Yanmei Liu*[2] *and Zilong Wen*[1]

[1]Department of Biochemistry, The Hong Kong University of Science and Technology, Clear Water Bay, Kowloon, P. R. China

[2]Max Planck Institute of Molecular Cell Biology and Genetics, Laboratory of Experimental Diabetology, Carl Gustav Carus Medical School, Dresden University of Technology, Dresden, Germany

The hematopoiesis in vertebrates includes two successive waves: 'primitive' and 'definitive' wave. Primitive hematopoiesis is a transient process and produces two blood cell lineages: myeloid (granulocytes and monocytes) and erythroid (erythrocytes and platelets) cells, whereas definitive wave, which is active throughout the lifetime of an organism, gives rise to all blood cells including erythrocytes, megakaryocytes (platelets), granulocyte, monocytes, and lymphocytes. Zebrafish has recently emerged as a powerful vertebrate model organism for study of hematopoiesis. Through forward genetic screen aimed to uncover zebrafish mutants with aberrant definitive hematopoiesis, a mutant zebrafish line Thy-193 has been identified.

By characterizing Thy-193 phenotype using whole mount in situ hybridization method, Thy-193 mutant shows normal primitive hematopoiesis, but aberrant maintenance of definitive hematopoietic stem cells(HSCs) without interfering the initiation of HSCs. By positional cloning, the Thy-193 mutant gene is mapped to sumolation activating enzyme subunit 1, sae1, which is a molecule involved in sumolation pathway. Sequencing of the sae1 gene shows a C817T mutation, resulting in a truncated protein lack of conserved C terminal 76 amino acids. This mapping result is further confirmed by rescue experiment with full-length sae1 RNA injection. Moreover, knocking down ubc9 to block sumolation pathway leads to a similar blood phenotype with Thy-193 mutant, implying sae1 might impact Thy-193 mutant via sumolation pathway.

* Email: xiuling@ust.hk.

Structural and Functional Analysis of Influenza H5N1 Nucleoprotein Reveals Amino Acids That Are Crucial for Oligomerization, RNA Binding and Ribonucleoprotein Activities

***A. K. L. Ng*[*1,2], *W. H. Chan* [1,2], *N. C. Robb* [4], *Lam MKH*[1], *P. K. S. Chan*[3], *Au SWN*[1,2], *E. Fodor*[4] *and P. C. Shaw*[1,2]**

[1]Department of Biochemistry and Centre for Protein Science and Crystallography,
[2]Molecular Biotechnology Programme
[3]Department of Microbiology, The Chinese University of Hong Kong,
Shatin, N.T., Hong Kong, China.
[4]Sir William Dunn School of Pathology, University of Oxford, South Parks Road,
Oxford, UK

The recent H1N1 influenza pandemic has become a major concern worldwide. The nucleoprotein (NP) of the virus binds the RNA genome and plays essential role in transcription and replication during the virus life cycle. We have determined the 3.3 Å crystal structure of H5N1 NP, which is composed of head and body domains and a tail loop. Using surface plasmon resonance, we found the basic loop (residues 73-91) and arginine-rich groove, but mostly a protruding element centering at R174 and R175, to be important in RNA binding by NP. To bind RNA, NP may first capture the RNA by the flexible basic loop and then the RNA is clamped by the protruding element. The crystal structure also revealed that NP forms homo-oligomers by inserting its distinctive tail loop into the body groove.

Using ribonucleoprotein (RNP) reconstitution assay, we identified E339, V408, P410, R416, L418 and L419 being crucial for forming a functional RNP, while R267, I406, R422 and E449 play supportive roles. Further characterization by static light scattering and co-immunoprecipitation showed that the reduction of RNP activities is due to the interference of NP homo-oligomerization.

We concluded that NP oligomerization event is governed by three factors: (1) interaction between the tail loop and the insertion groove; (2) maintenance of the tail loop conformation; and (3) stabilization of NP homo-oligomer. Our study of H5N1 NP sheds light on inhibitor design for combating influenza virus.

* Email: ng0567@gmail.com.

Structural Basis of the Organization of the USH1 Interactome

Lifeng Pan*, Jing Yan, Lin Wu and Mingjie Zhang
Department of Biochemistry, Hong Kong University of Science and Technology, Clear Water Bay, Kowloon, Hong Kong, China

The hereditary hearing-vision loss disease Usher syndrome I (USH1) is caused by defects in several proteins, which form an integrated protein network called the USH1 interactome. The proper assembly of the USH1 interactome is essential for both the development and maintenance of visual and hair cells. However, the biochemical and structural mechanism governing the USH1 protein complex formation is largely unclear. Using NMR spectroscopy, we first elucidate the multi-dentate binding mode between harmonin and cadherin23, which provides a structural and biochemical basis for the harmonin-mediated assembly of stable tip link complex in the auditory hair cells. Then using X-ray Crystallography combined with NMR, we solved the high resolution structure of the nucleation core of the USH1 complex formed by harmonin and Sans. The complex structure not only reveals an unexpected assembly mode between harmonin and Sans but also uncovers a previously unknown interaction mode for both PDZ and SAM domains. The synergistic PDZ1/SAM and PDZ1/carboxyl PDZ binding-motif interactions between harmonin and Sans lock the two scaffold proteins into a highly stable complex, acting as the nucleation core of the USH1 interactome. In summary, our findings provide the first atomic picture showing the assembly of the USH1 interactome.

In addition to providing molecular explanations to the disease-causing mutations already found in USH1 patients, the complex structures will undoubtedly be valuable in guiding future discoveries of other mutations in these USH1 proteins in human patients.

* E-mail: panly@ust.hk.

Index

A

B

C

D

E

F

G

H

I

N

O

P

Q

R

S

T

U

V

W

X

Y